EAI/Springer Innovations in Communication and Computing

Series Editor

Imrich Chlamtac
European Alliance for Innovation
Ghent, Belgium

Editor's Note

The impact of information technologies is creating a new world yet not fully understood. The extent and speed of economic, life style and social changes already perceived in everyday life is hard to estimate without understanding the technological driving forces behind it. This series presents contributed volumes featuring the latest research and development in the various information engineering technologies that play a key role in this process.

The range of topics, focusing primarily on communications and computing engineering include, but are not limited to, wireless networks; mobile communication; design and learning; gaming; interaction; e-health and pervasive healthcare; energy management; smart grids; internet of things; cognitive radio networks; computation; cloud computing; ubiquitous connectivity, and in mode general smart living, smart cities, Internet of Things and more. The series publishes a combination of expanded papers selected from hosted and sponsored European Alliance for Innovation (EAI) conferences that present cutting edge, global research as well as provide new perspectives on traditional related engineering fields. This content, complemented with open calls for contribution of book titles and individual chapters, together maintain Springer's and EAI's high standards of academic excellence. The audience for the books consists of researchers, industry professionals, advanced level students as well as practitioners in related fields of activity include information and communication specialists, security experts, economists, urban planners, doctors, and in general representatives in all those walks of life affected ad contributing to the information revolution.

Indexing: This series is indexed in Scopus, Ei Compendex, and zbMATH.

About EAI

EAI is a grassroots member organization initiated through cooperation between businesses, public, private and government organizations to address the global challenges of Europe's future competitiveness and link the European Research community with its counterparts around the globe. EAI reaches out to hundreds of thousands of individual subscribers on all continents and collaborates with an institutional member base including Fortune 500 companies, government organizations, and educational institutions, provide a free research and innovation platform.

Through its open free membership model EAI promotes a new research and innovation culture based on collaboration, connectivity and recognition of excellence by community.

More information about this series at http://www.springer.com/series/15427

Sasmita Rani Samanta • Pradeep Kumar Mallick
Prasant Kumar Pattnaik • Jnyana Ranjan Mohanty
Zdzislaw Polkowski
Editors

Cognitive Computing for Risk Management

Springer

EAI
RESEARCH MEETS INNOVATION

Editors
Sasmita Rani Samanta
Professor of Management
Pro-Vice Chancellor
Kalinga Institute of Industrial
Technology (KIIT)
Bhubaneswar, Odisha, India

Prasant Kumar Pattnaik
School of Computer Engineering
Kalinga Institute of Industrial Technology
Bhubaneswar, Odisha, India

Zdzislaw Polkowski
Department of Business Intelligence
in Management
Wroclaw University of Economics
and Business
Polkowice, Poland

Pradeep Kumar Mallick
School of Computer Engineering
Kalinga Institute of Industrial Technology
Bhubaneswar, Odisha, India

Jnyana Ranjan Mohanty
School of Computer Applications
Kalinga Institute of Industrial Technology
Bhubaneswar, Odisha, India

ISSN 2522-8595 ISSN 2522-8609 (electronic)
EAI/Springer Innovations in Communication and Computing
ISBN 978-3-030-74516-5 ISBN 978-3-030-74517-2 (eBook)
https://doi.org/10.1007/978-3-030-74517-2

This Springer imprint is published by the registered company Springer Nature Switzerland AG
The registered company address is: Gewerbestrasse 11, 6330 Cham, Switzerland

Preface

This edited book aims to bring together leading researchers, academic scientists, and research scholars to put forward and share their experiences and research results on all aspects of cognitive computing and risk management. It also provides a premier interdisciplinary platform for educators, practitioners, and researchers to present and discuss the most recent innovations, trends, and concerns as well as practical challenges encountered and solutions adopted in the field of cognitive computing. The book is organized into ten chapters,

Chapter 1 deals with review research that tries to assess the broad direction in entrepreneurship cognition context, analyzes various themes from the published articles, and critically analyzes the course.

Chapter 2 focuses on exploring the relationship between select cognitive distortions and stress in a college setting. Partly modified scales of cognitive distortions (Roberts, M. B) and emotional stress (depression anxiety stress scales (DASS) have been used for the survey). Results indicate that there is a moderate correlation between the cognitive distortions and emotional stress. Results may be useful to mitigate the negative consequences of emotional stress born of cognitive distortions. It is implied that teachers recognize the suffering of general students in this respect and must adopt useful techniques to help the students.

Chapter 3 deals with the objective of the study, which is to find the influence of cognitive barriers on drug adherence. Three cognitive barriers, viz., denial, refusal, and mistrust, have been taken in the study, and their effect on the drug adherence of diabetic patients have been studied.

Chapter 4 analyzes the effect of the founding ecosystem components on overcoming the death-valley of start-up companies through the mediating effect of establishing a win-win platform in Korea. The study has shown that the start-up ecosystem has a significant impact on overcoming start-up death valleys, and, in particular, research on win-win platforms and optimal configuration scenarios is needed. We will pursue sustainable management of Startup and will soon be the birth of The Decacorn SME company in Korea by this chapter.

Chapter 5 clarifies the mediating effect of the causal relationship and relationship commitment that the collaboration motive of the partners participating in the consulting project has on the collaboration performance. This study is by a consultant who conducts consulting on SMB. Data was collected using an online questionnaire.

Chapter 6 presents about the research is to search out how the consulting performance characteristics, such as the expertise of consultants, the fidelity of the support of the government agencies, and the acceptance attitude of companies, affect the quality of consulting results when SMEs using the government-supported consulting program receive consulting. In addition, to search whether there is an influence on the application of consulting results in their work according to consulting performance characteristics.

Chapter 7 presents an approach that based on previous studies that show that the higher a company's sustainable management activities are, its management performance changes. The purpose is to examine the impact of corporate sustainable management activities on corporate management performance and to verify the mediating role of competitive advantage.

Chapter 8 explores the research targets, including the current consultants who are at present conducting consulting, and students and graduates of consulting graduate schools. Data was collected through a questionnaire. This research is an empirical proof analysis of the impact of SMEs' security culture formation on the security system and sustainable growth. We then performed an analysis of the effectiveness of the security system.

Chapter 9 focuses on identifying the degree of increase in the internal effects of SMEs employees of primary business partners of large enterprises through supporting SMEs of large enterprises for win-win growth. This study was conducted based on basic data obtained through a web survey of employees of primary partner of large enterprises.

Chapter 10 aims to examine the relationship between the competency of consultants, the key factors in management consulting, the quality of SME management consulting services, and, finally, the management performance of SME. This study verified the critical qualities of consultants and quality factors of consulting services that SEMs perceived as necessary, and also revealed the relationship with management performance.

We are very thankful to Almighty to supporting and standing at all times with us, whether it's good or tough times and given ways to concede us. Starting from the Call for chapters till the finalization of chapters, all the editors given their contributions amicably, which is a positive sign of significant team works, The editors are sincerely thankful to all the members of Springer especially Eliška Vlčková,

Managing Editor, European Alliance for Innovation (EAI), Chlamtac Imrich, Series Editor and Mary James, Senior Editor for providing constructive inputs and allowing an opportunity to edit this important book. We are equally thankful to reviewers who are from different places in and around the globe shared their support and stand firm towards quality chapter submissions.

Bhubaneswar, Odisha, India — Sasmita Rani Samanta
Bhubaneswar, Odisha, India — Pradeep Kumar Mallick
Bhubaneswar, Odisha, India — Prasant Kumar Pattnaik
Bhubaneswar, Odisha, India — Jnyana Ranjan Mohanty
Polkowice, Poland — Zdzislaw Polokowski

Key Features

1. Addresses the issues in cognitive computing.
2. Addresses the complete functional framework workflow in cognitive computing
3. Explores basic and high-level concepts, thus serving as a manual for those in the industry while also helping beginners to understand both basic and advanced aspects in risk management–related issues.
4. Bases itself on the latest technologies, and covers the major challenges, issues, and advances in cognitive computing.
5. Applies cognitive computing to the risk domain
6. Explores other opportunities for cognitive computing through cognitive computing and its implications to the real world.
7. Augments and empowers business decisions

About the Book

Management activities in cognitive management are carried out via tools affecting human cognitive capabilities. The subject field of cognitive management is the process of managing organizational knowledge, which is possible in the information society and is most effective in a social environment. Applying cognitive computing against massive data sets can help organizations process information more quickly and make smarter business decisions. And cognitive computing is increasingly being used in the domain of risk management, mining often ambiguous and uncertain data to find indicators of known and unknown risks. This edited book discusses applications of cognitive computing for risk management.

Contents

About the Editors

Sasmita Rani Samanta, Ph.D. (Management) fellow of the Royal Society of Arts (FRSA, UK), fellow of the Computer Society of India (FCSI), currently serving as a full professor of commerce and management, and pro vice chancellor, KIIT Deemed to be University, India. She has received 25 of awards from different national and international organizations for her outstanding contribution to the field of education and leadership.

Pradeep Kumar Mallick, Ph.D. (Computer Science) is currently working as associate professor in the School of Computer Engineering, Kalinga Institute of Industrial Technology (KIIT) Deemed to be University, Odisha, India. He has completed his postdoctoral fellowship (PDF) at Kongju National University, South Korea; Ph.D. at Siksha 'O' Anusandhan University; M.Tech. (CSE) from Biju Patnaik University of Technology (BPUT); and M.C.A. from Fakir Mohan University, Balasore, India. Besides academics, he is also involved in various administrative activities, Member of Board of Studies, Member of Doctoral Research Evaluation Committee, Admission Committee, etc. His areas of research include algorithm design and analysis as well as data mining, image processing, soft computing, and machine learning. He has published several book chapters and papers in national and international journals and conference proceedings.

Prasant Kumar Pattnaik, Ph.D. (Computer Science) fellow of the IETE, senior member of the IEEE, is a professor in the School of Computer Engineering at KIIT Deemed University, Bhubaneswar. He has more than a decade of teaching and research experience. Dr. Pattnaik has published several of research papers in peer-reviewed international journals and conferences. He has also published many edited book volumes with Springer, IGI Global, and Wiley Publication. His areas of interest include mobile computing, cloud computing, cyber security, intelligent systems, and brain–computer interface.

Jnyana Ranjan Mohanty, Ph.D. (Computer Science) is currently working at KIIT University, Bhubaneswar, Odisha, India, as a professor and registrar. He earned his Ph.D. degree in computer science from Utkal University, Bhubaneswar, in the year 2008.He is a life member of the Indian Society for Technical Education (ISTE). He is actively engaged in research work. His research interests include queuing theory, computational intelligence, and cloud computing.

Zdzislaw Polkowski, Ph.D. is professor of UJW in the Faculty of Technical Sciences and Rector's Representative for International Cooperation and Erasmus+ Program at the Jan Wyzykowski University Polkowice. Since 2019, he is also adjunct professor in the Department of Business Intelligence in Management, Wroclaw University of Economics and Business, Poland. He is the former dean of the Technical Sciences Faculty during the period 2009–2012 at UZZM in Lubin. He holds a Ph.D. degree in computer science and management from Wroclaw University of Technology, a postgraduate degree in microcomputer systems in management from the University of Economics in Wroclaw, and a postgraduate degree IT in education from Economics University in Katowice. He obtained his engineering degree in computer systems in industry from the Technical University of Zielona Gora, funded by the European Commission. He also owns an IT SME consultancy company in Polkowice and Lubin, Poland.

The Cognitive Approach to Entrepreneurship: An Agenda for Future Research

Brajaballav Kar, Nilamadhab Kar, and Chandrabhanu Das

1 Introduction

Entrepreneurship is an intriguing human endeavor undertaken in a specific context that has eluded a comprehensive understanding so far. There are considerable complexities in the entrepreneurial processes. Compared to other branches of social sciences, entrepreneurship is still evolving, borrowing substantive concepts from different parent branches. Entrepreneurship research, as discussed in later parts, is yet to contribute substantially to other branches of social science. The initial stages of investigation in this field focused on the entrepreneur as a distinct individual with identifiable traits and behavior. Subsequent stage focused on the influence of multiple contextual variables and processes involved in entrepreneurship. The entrepreneurial personality-based research could not substantially distinguish the entrepreneurial processes [1].

The article "Cognition, Choice and Uncertainty" by Buchanan and Di Pierro [2] is one of the initial articles, and it criticized the theory of choice under uncertainty due to its neglect of the role of cognition. It also discussed the role of information in reducing cognitive uncertainty in an entrepreneurial context [2]. Research linking cognition to entrepreneurship has gained momentum, starting in early 2000, investigating the entrepreneurial mindset, thinking, decision-making, and various other factors related to the decision-making processes. Primarily, researchers claim that cognitive research has gained momentum due to the limitations of the trait approach. Broadly, cognitive research is trying to link and explain entrepreneurial behavior

B. Kar (✉) · C. Das
School of Management, KIIT University, Bhubaneshwar, India
e-mail: brajkar@gmail.com

N. Kar
Black Country Healthcare NHS Foundation Trust, Wolverhampton, UK

S. R. Samanta et al. (eds.), *Cognitive Computing for Risk Management*, EAI/Springer Innovations in Communication and Computing, https://doi.org/10.1007/978-3-030-74517-2_1

using cognitions. The thrust areas of cognitive research study elements such as scripts, cognitive models of choice, information processing, entrepreneurial learning, self-efficacy, cognitive styles, risk perception, and heuristics at the individual and group level [3]. The evolving subdomain of cognitive research in entrepreneurship is undergoing various attempts to define, clarify, and guide the scope of research.

In this article, we try to understand the broad direction of extant cognitive research in entrepreneurship research by thematic groupings of articles obtained through a structured literature review. We discuss cognition and social cognition definition in the next section. Subsequently, we discuss the two-staged review study design adopted for this chapter, major themes from the existing research to understand the broad direction. The discussion section categorizes themes according to the accepted categorization of cognition. Various data sets and methodologies used in extant research are also indicated. We have improved upon the structured literature search by a second-stage purposive literature inclusion and review. Finally, we present some directions for future research.

2 Cognition and Social Cognition

Cognition is generally understood as mental processes in developing comprehension and knowledge. Language, attention, gathering information, remembering, thinking, learning, problem-solving, perception, and judging are some of the processes involved in cognition. A broad definition of cognition was given by Neisser [4] as "all processes, by which sensory input is transformed, reduced, elaborated, stored, recovered, and used [4]." Beyond the broad definition of cognition, there are many theoretical bases, on which entrepreneurship research depends. One of the most used theoretical frameworks in entrepreneurship research is Albert Bandura's social cognitive theory (SCT) that was an offshoot of his social learning theory. It posits that the person, environment, and behavior interact and influence each other dynamically and reciprocally. Six constructs such as reciprocal determinism, behavior capability, observational learning, reinforcements, expectations, and self-efficacy were used to model the interactions. Some criticisms include that the theory assumes the changes to be automatic, does not consider the relative importance of factors, and does not include emotion or motivation. Despite criticisms, it is one of the most popular theories used in entrepreneurship research.

Social cognition indicates that the response to a situation is an interaction of the perception of self and others and the interpersonal knowledge. The processes of interaction may be implicit or explicit and are subject to bias [5]. The entrepreneurial decision-making uses combinations of cognitive factors termed as cognitive map. One of the typologies of cognitive maps suggests six categories such as purpose-driven, determined and hesitant, value-based, vacillating and unintended, based on single motive, or based on single solution [6].

The cognitive approach to entrepreneurship research is evolving, and hence a stocktaking of the extant research is essential to understand the broader direction and effort by researchers. Secondly, a structured review finds issues for subsequent research focus. We have taken these two primary objectives to assess the overall research direction and indicate gaps, issues, and challenges involved in the cognitive approach to entrepreneurship research. The following section discusses the review study design.

3 Literature Review Study Design

The review objective, in this case, is to summarize the large field of research, and hence a narrative review approach was considered appropriate [7]. The second objective was to identify areas intersecting with entrepreneurship and cognition. We considered a mixed literature review design; at the first stage, the literature was searched from Scopus and Web of Science databases, in a structured literature review method, and subsequently, further literature was added to consolidate the literature from other electronic databases or search engines (Google Scholar) on the theme derived from the structured literature review.

For the first stage of the search, we included journal articles in the English language without mentioning the time until October 2020 and excluded books and conference proceeding publications. The relevance and quality were assessed from the abstract and publication source (Table 1).

Academic publisher of repute is one of the quality indications of the articles chosen for review. In this structured review, the following publishers and number of articles were included: Elsevier (20%), Emerald (17%), Sage (16%), Springer

Table 1 Search details of articles, inclusion and exclusion criteria

Database and query	Constraints (inclusion/exclusion criteria)	Records	Duplicates and irrelevant to the topic	Final articles reviewed
Scopus query: ITLE-ABS-KEY ("Cognitive Theory" entrepreneurship) AND (LIMIT-TO (LANGUAGE, "English")) AND (LIMIT-TO (SRCTYPE, "j"))	Search inclusion criteria, language English, type of document – journals (books, conference paper excluded); excluded	49	Removed = duplicates (38) + irrelevant (50)	88
Web of Science query: ALL FIELDS: ("Cognitive Theory" Entrepreneurship) Timespan: All years. Indexes: SCI-EXPANDED, SSCI, A&HCI, CPCI-S, BKCI-S, ESCI		127		

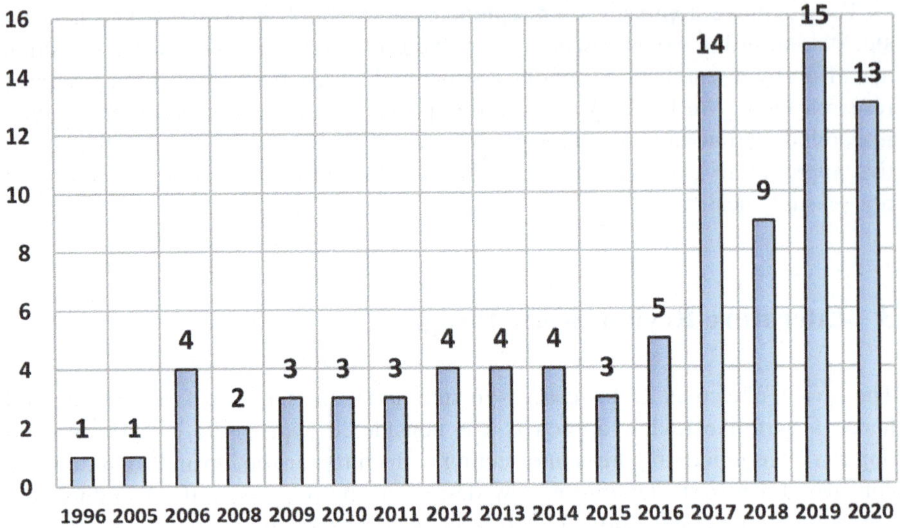

Fig. 1 Distribution of year and number of articles

(14%), Wiley (7%), Frontiers (5%), Inderscience, World Scientific and Taylor and Francis (3% each), and MDPI (2%). Cumulatively, these were 91 percent of the total articles. The articles from the rest of the publishers (De Gruyter, AOM, PLOS, and others) were 1 percent each.

The journals were mostly subject-specific. The distribution of article according to journal focus was as follows: entrepreneurship (52%), psychology (16%), business and management (including technology, innovation, design, and others) (23%), education and career (8%), and health (1%). The detail of journals and the number of articles are given in appendix.

Figure 1 presents the distribution of the year and number of articles selected in this study. The cognition-related entrepreneurship research has increased substantially over the period.

3.1 Theoretical Foundations in the Selected Studies

As is known, entrepreneurship literature is multidisciplinary in nature. In this review, the researchers have studied different theoretical perspectives. In this section, we present various theories, principles, and perspectives used in the selected literature. The cognitive theory was a search criterion for the literature, and unsurprisingly many literatures in this study mention the social cognitive theory [8–16]. Ajzen's theory of planned behavior (TPB) is the second most referred theory in the selected research literature [11, 17–20]. The variants of SCT such as entrepreneur cognitive theory have also formed the base of some researches [21]. Researchers have also tried to propose a theory of international entrepreneurship [22],

Learning- and career-related theories have influenced the research focus. The social learning theory [9], social career cognitive theory [23–25], entrepreneurship career development theory [11], and career construction theory [26] have been identified in this review.

The information processing aspect in entrepreneurship has also been based upon theoretical frameworks such as the information-processing framework of entrepreneurship (I-Pe) [27] and affect-as-information theory [28]. The comparative-judgment-formation framework [29] and affective processing principle [28] are used in the studies as well.

Entrepreneurship is a social phenomenon, and the theory of social capital remains the predominant paradigm [21]. The organizational aspects of a new venture have considered the stakeholder theory as the base [30]. The ethical aspects are based upon the moral self-regulation theory ([16, 31]). The psychological theories on grief were used to understand how grief influences subsequent entrepreneurship attempts [32]. The risk aspects involved in entrepreneurship have used prospect theory [33]. The stereotype activation theory has been used to understand prejudices prevailing among entrepreneurs [34]. The egocentrism theory [29] and goal theory [35] have focused on the cognitive aspects of entrepreneurs. The institutional theory explicates the role of institutions in influencing various aspects of entrepreneurship processes [23]. The theory of effectuation has been used in studies of Stroe et al. [36] and Zhang et al. [37].

The literature review also indicates a variety of concepts on which researches have been conducted. The majority of literature focused on self-efficacy, specifically entrepreneurial self-efficacy (ESE) and its variants like creative self-efficacy. Entrepreneurial traits, intention, motivations, leadership skills, attention, passion, ethics, and self-esteem are some of the individual factors that have been considered in the research framework. Career switch from veterans to entrepreneurship has been investigated in one of the researches included in this study. A gender-based research investigates the role of domestic violence on women entrepreneurship. Similarly, emotions such as fear, joy, and anger experienced by entrepreneurs during different phases of the entrepreneurial journey also have been considered. Some higher-order skills such as social competence, resilience, collaborative behavior, cognitive biases (example overconfidence), creativity, values (independence, creativity, ambition, and daring), emotional intelligence, dynamic capability, growth mindset, and sense-making are also actively pursued in the entrepreneurship literature. Many traits are clubbed under socio-cognitive traits as a broader category.

Constructs, closely identifiable with entrepreneurship literature, such as opportunity recognition, innovation, risk, decision-making, and self-employment, remain in research focus. Researchers have also investigated the technology adoption model (TAM). The role of social capital, social environment, social media, networks, family business, succession, and mentoring continues to be investigated. Similarly, learning, information exchange, and locus of control are being investigated as entrepreneurship processes. The role of the entrepreneurial ecosystem and institutional context on entrepreneurship is actively pursued. Intrapreneurship and social entrepreneurship continue to be sub-segment of entrepreneurship research. Immigrant entrepreneurship and its related development are the emerging areas in entrepreneurship research. Entrepreneur well-being also seems to be an emerging area.

4 Discussion on Research Themes

For the categorization of the themes, we considered the broad definition of cognition and social cognitive research constructs. We also considered categories such as person, situation, cognition, and motivation as defined by Crocker et al. [38]. The following were the final categories and sub-categories identified in this study: personal factors (age, gender, education, network), entrepreneurial qualities (intention, social cognition, creativity, self-efficacy, leadership skill, values ethics, resilience, growth mindset, emotions, and well-being). Family business and intrapreneurship have been considered as separate themes because of their distinctiveness of the entrepreneurial process. Similarly, the entrepreneurial process factors such as entrepreneurial opportunity, information processing, and decision-making-related researches are considered as specific categories each. The situation variables such as geography and sector are identified as separate categories. Each research explores multiple constructs and reports their findings. We have used their major findings for categorization to a specific theme.

4.1 Personal Factors

Demography factors are important independent variables with a possible influence on various other aspects of entrepreneurship. The variables include gender, age, race, marital status, education, income, and family size, among others. Gender studies on entrepreneurship are also grouped under this theme.

4.1.1 Age

Entrepreneurship can be undertaken at any age. The age-specific characteristics like experience, maturity, and dynamism are likely to be different. The evidence is not conclusive, indicating a possible interaction effect of larger contextual variables.

The social learning theory and the social cognitive career theory have been used to understand the interest in self-employment of older workers. Attitudes toward aging, anticipated longevity, and support from network predicted entrepreneurial self-efficacy and expected outcome. The relationship between social support and entrepreneurial interest was mediated by the outcome expectations. The age-related psychological, social, and cultural factors, as well as the contextual environment, influence the interest in entrepreneurship [9]. Similarly, veterans usually acquire leadership skills and understanding of technology from their prior experience. In their case too, the entrepreneurial intention is positively influenced by their self-efficacy, personal attitude and beliefs, tolerance of ambiguity, and risk-taking propensity [39].

The performance of the firm is one of the important criteria evaluated against demographic characteristics. It can be the financial performance (objective) or a set of parameters accepted by the entrepreneur (subjective). The subjective performance does not depend on the demographic characteristics of entrepreneurs (age, experience, and education) [21]. However, the network characteristics such as size, density, centrality, competitive or supportive network, and self-efficacy explain the subjective performance [21]. Authors have explained it based on social capital and entrepreneur cognitive theory.

4.1.2 Gender

Women's participation by number or proportion is much less, even in advanced economies. The influence of other socioeconomic and cultural variables becomes inherent in the gendered perspective. It is more likely that men in patriarchal societies and women in matrilineal society attempt a new venture. The ESE, and fear of business failure, in both societies, explains this gender gap. Thus, the gender role has a social context, and it is argued that the entrepreneurial propensity is due to the outcome of socialization [40]. A gender-based stereotype is associated with women's entrepreneurial intention and self-efficacy with high-growth entrepreneurship that limits their behavior. The stereotype acts as a barrier to entrepreneurship and needs to be decreased [34]. The role of family support and direct involvement along with the perception about one's ability positively moderates the relationship between the attitude and intention toward growth [41].

Domestic violence in form of exposure to physical or sexual abuse inhibits the initiation of new business even if the women have access to financing. Such victims are less likely to attempt a new business compared to others without any such adverse experience. Authors have explained that it is due to the reduction of ESE and an increased fear of business failure. Exposure to environmental disasters or a traumatic event also reduces the intention to initiate a new business [42].

4.1.3 Role of Education

Entrepreneurship education has increased over a period. However, the influence of education on various phases of the entrepreneurial process is not understood comprehensively. Secondly, samples from the participants in entrepreneurship education for the study introduce some degree of bias. Thirdly, such education has concomitant learning that needs focused investigation. Authors have indicated that the role of entrepreneurial education on entrepreneurial behavior is ambiguous and some emotions of individuals may be the missing link [43]. Indicating that there is a lack of planning in the current entrepreneurship education for the creation of new business, authors have taken a normative approach to suggest that the educators should enhance students' intentions and self-perceptions. They have suggested that entrepreneurial cognition and risk may provide a theoretical perspective to develop analysis and intuition as modes of thinking. Education should develop entrepreneurial intention and passage to action [44]. Based on social cognition theory, and social constructivism, researchers also have argued about the need for creativity, innovation, and entrepreneurship as meta-competencies that need to be developed by the education system for entrepreneurial culture. Depending on the paradigm of entrepreneurship education, the objective or outcome changes. For example, it can be education for, about, or through entrepreneurship. It is important to underscore that students of such a program consider themselves creative and are ambivalent on the role of education [45].

Other studies linking education and entrepreneurial intention have reported the mediating role of ESE and passion under a boundary condition of team cooperation [16]. It is also reported that the ESE drives the classroom activity that enhances intention, but common activities generally found in entrepreneurship education course negatively impact ESE [46]. Gender acts as a moderator in the role of education in enhancing the ESE, and thus the education should have appropriate gender-specific interventions [47].

Researchers have also tried to investigate if entrepreneurship education enhances entrepreneurial activities across national cultures. They based their study on the SCT and claim that entrepreneurship education becomes more important for entrepreneurial activities in countries with higher individualism, reduced uncertainty avoidance, and higher levels of masculinity. They extended their findings to international entrepreneurship arguing it to depend upon national culture [48].

Other sets of literatures focus on the learning, mentoring, and role model aspects of education. A referent individual is also taken to be a role model. An experimental research found that entrepreneurial role models have less impact on the students with prior entrepreneurial experience, higher self-esteem, and internal locus of control [49]. Similarly, mentoring is assumed to provide support to novice entrepreneurs to gain confidence and achieve the desired outcomes. However, mentoring was found to effect the achievement of outcomes such as satisfaction, need for autonomy, and intention to stay self-employed through enhanced self-efficacy [50]. Research also reported that the entrepreneurial inclination and learning tendency have helped in the process of technology adoption [51].

4.1.4 Network

Since the psychological variables, personality traits, and demographic factors, in an isolated way, have not been able to differentiate entrepreneurial behavior, researchers attempted to model possible interactions of these variables to improve the predictability. Researchers extended the opportunity and individual nexus to social capital and cognition to explain why only a few exploit an opportunity [52]. The network characteristics such as strength and diversity are believed to influence the entrepreneurial process. The network embeddedness was reported to influence intention [15]. The network is also exploited for marketing in the case of immigrant entrepreneurs [10]. Similarly, a study found that the competitive and supportive network predicted subjective performance significantly [21].

4.2 Entrepreneurial Qualities

Of many qualities, few are believed to be strongly related to entrepreneurship. Qualities like risk tolerance, passion, proactiveness, optimism, achievement orientation, creativity, narcissism, and others are actively used in entrepreneurship research. These qualities are grouped to a second-order level, e.g., entrepreneurial orientation to understand the influence on each other and on qualities that are mostly accepted to influence entrepreneurial processes.

Constructs from goal theory and SCT have been used to understand the motivation of entrepreneurs to exploit opportunities and outcomes of a new venture. It is also used to understand the likelihood of quitting or continuing the nascent start-up activities [35]. The concepts of individual qualities are extended to organizational qualities since an entrepreneurial organization is most likely to inherit such qualities from the founding entrepreneur. The values, attitudes, and self-efficacy are likely to be inherited and subsequently influence the firm's entrepreneurial and market orientation leading to firm performance. Studies have indicated a relationship between the entrepreneurs and their firm strategies and firm performance, at least at the early stages. The internal values of the firm were found to be positively related to entrepreneurial attitude. The attitude, self-efficacy, and innovativeness were positively related. The market orientation of the firm was related to the proactiveness and entrepreneurial attitude. Finally, the firm performance was related to the market orientation, self-efficacy, and innovativeness [53].

4.2.1 Intention

The intention is one of the most frequently studied constructs in entrepreneurship research. The entrepreneurial intention is influenced by entrepreneurial qualities and knowledge, perceived entrepreneurialism, and entrepreneurial inspiration (EIN). The findings have shown that self-efficacy partially mediates the relationship of entrepreneurial qualities, perceived entrepreneurialism, and EIN on intention. The efficacy, however, acts as a mediator in the relation between entrepreneurial knowledge and intention [54].

The self-regulation perspective indicates that individuals control and evaluate their own learning. Such a perspective has been used to model the risk propensity, intention, and self-efficacy. The ESE and the regulatory focus of promotion influence intention positively. However, the association between risk propensity and intention was found to be partially mediated by the promotion and prevention focus of self-regulation [55].

Self-beliefs like general and entrepreneurial self-efficacy and global self-esteem seem to strongly influence entrepreneurial intention and actual business start-up. The general and entrepreneurial self-efficacy beliefs were found as key predictors of intention [56].

Many entrepreneurs take up entrepreneurship driven by their hobby. Entrepreneurial passion and creativity are suggested to be the antecedents of entrepreneurial intentions. Entrepreneurial passion was found to have a strong positive relationship with intentions, even when ESE was introduced as a mediator. Self-efficacy also mediated the relationship between creativity and entrepreneurial intentions. The study proposed that individuals need to feel self-efficacious enough to take up entrepreneurship [57]. Based on the SCT, another study proposed that creativity should be in the models of intention [58]. A study based on TPB showed that the role models have an indirect influence on intention through its antecedents. The relationship between perceived behavior control and intentions did not indicate any gender difference, though the other relationships in TPB were moderated by gender. The entrepreneurial intentions were predicted by subjective norms and attitude toward entrepreneurship [18].

What additional factors influence the entrepreneurial intention of the students? This question was addressed in the study by extending the TPB framework to include perceived career options, entrepreneurial knowledge, and traits. The availability of a viable career option and entrepreneurial knowledge and positive attitude toward entrepreneurship was found to influence intention significantly. Subjective norms had a weak influence on intention. As found in most of the studies, the link between trait and intention acts through the mechanism of perceived self-efficacy [19].

The interaction between cognitive factors and environmental munificence also influences entrepreneurial intentions. The munificence of the environment becomes a contingent factor in an incubator setting, by strengthening the relationship between risk-taking propensity and ESE [8].

The emphasis on social entrepreneurship has led some researchers to investigate a distinct social entrepreneurial intention. Using the SCT perspective, researchers tried to understand the social entrepreneurial intention, network embeddedness, and opportunity identification process in entrepreneurship [15]. The social entrepreneurial intention is found to be influenced by emotional intelligence, social work experience, previous entrepreneurial experience, and environmental support. The ESE mediates the relationship with other factors. Environmental support has no significant relationship with entrepreneurial intention [59].

The reentry intention indicates the entrepreneur's intention to start again following a business exit. It is a puzzling observation that many failed entrepreneurs reenter. The prospect theory and self-efficacy may predict opposite actions in this context. Researchers have theorized a moderating model with boundary conditions.

They argue that the reentry is explained by prospect theory when the business exit was due to moderate or low self-efficacy [33].

4.2.2 Social Cognition

Processes associated with venture formation process are a social process, and social cognition of entrepreneur becomes crucial in venture formation and growth. Unsurprisingly, it is one of the most frequently used bases for entrepreneurship research.

There are various decision-making styles, different levels of environmental uncertainties, and relations between the organizations, which influence the social cognition aspects. It is found that increased market uncertainty rushes the entrepreneurial decision-making process. The overall decision-making process also has an intuitive and rational dimension [60]. Studies also indicate that the demographic characteristics (gender and age), social capital, as well as dispositional regulatory focus impacted the evaluation decisions [12]. The SCT formed the basis of study in most research.

4.2.3 Creativity and Innovation Ability

Creativity or innovation ability is one of the hallmarks of entrepreneurship. Individual and group fit (I-G fit) has an impact on innovation; entrepreneurial cognition, innovation, ability, and environment act as incentives through a sense of psychological ownership [61]. The new product development (NPD) process and entrepreneurship draw on the SCT and behavioral integration research. Researchers studied the influence of collaborative behavior, information exchange, joint decision-making, and collective efficacy interactions on innovation performance. They reported information exchange to be positively associated with innovation performance. The innovation performance was positively but marginally influenced by collaborative behavior but, interestingly, was not influenced by joint decision-making. Collective efficacy had a moderating effect. When the collective efficacy was high, both collaborative behavior and joint decision-making influenced innovation performance to a higher degree. Similarly, with higher collective efficacy, information exchange influenced innovation performance less positively. The research argued collective efficacy, as a limiting condition for the integration of behavioral dimensions [62].

Several entrepreneurship studies investigated the determinants of innovation outcomes in small and medium enterprises (SMEs). Even though entrepreneurial creativity is considered as a precondition, some argue that it is not the sole determinant of innovation. Models based on the social cognitive and innovation theory included entrepreneurial creativity, self-efficacy, as well as innovation outcomes and reported partial moderation effects of ESE [63].

Entrepreneurship in the technology domain is vital but difficult in mature industries because many cues confound the requirement for change, or effective interpretation of new data is necessary. Contrary to the popular belief, researchers indicate that technology entrepreneurship is typically a social activity, which requires joint efforts to reinterpret ambiguous data. A joint understanding is necessary for sustained technology efforts and a persistent, coordinated effort to achieve technological change. Sense-making, alertness to anomalies, and joint creation of an innovative as well as a shared cognitive context are some of the characteristics of technological entrepreneurship [64]. The technology adoption behavior of innovation was found to be influenced by the entrepreneurial inclination, ethical proclivity, marketing, self-efficacy, and outcome expectancy. The ethical tendencies and technology marketing indicated the behavioral intention to adopt new technological resources [65].

It is observed in almost all economies that women are under-represented in entrepreneurship. This prompts the question on the gender difference in creativity as a distinct dimension of entrepreneurship. There seems to be a gender dimension to creativity, problem management, risk management, and self-confidence, where only the self-confidence is at the same level between genders [66].

4.2.4 Entrepreneurial Self-Efficacy (ESE)

Self-efficacy is one of the dimensions of social cognitive theory. Though entrepreneurial self-efficacy is widely used in research, researchers have sought to clarify its dimensionality. One review proposed three dimensions of entrepreneurial self-efficacy as (1) particular aspect of entrepreneurship to which self-efficacy is applied (business start-up or growth activities), (2) the content of self-efficacy beliefs (task belief or outcome goal beliefs), and (3) the valence of entrepreneurial self-efficacy beliefs (positive or negative control beliefs) [67].

Deriving from social cognition, authors have argued self-efficacy is the cognitive representation of past success and beliefs and is a cognitive motivational resource [68]. In many studies, ESE is usually modeled as a mediator between different constructs. Self-efficacy is perceived to be related to entrepreneurial identity and is a schema of the individual as an entrepreneur. Authors have found that the ESE and entrepreneurial alertness predicted the entrepreneurial intention in the case of students and argued it to be taken as cognitive predictors [69].

Higher self-efficacy is associated with an increased likelihood of becoming a nascent entrepreneur, investing in entrepreneurial opportunities more aggressively, and creating and operating a business. Further, it increases the share of personal wealth invested in the business and the time duration spent in their own venture [70]. A longitudinal study based on SCT in real-life settings tested and confirmed the reciprocal relationships existing between self-efficacy, positive affect, and work engagement among entrepreneurs. Enthusiasm predicted work-related self-efficacy and work engagement in entrepreneurs. Self-efficacy beliefs predicted work engagement and enthusiasm, while work engagement predicted enthusiasm [13].

Various models based on constructs such as outcome expectation, subjective norms, ESE, and intentions have been investigated. The intention and ESE were found to be mediated by outcome expectation but moderated by subjective norms [24]. The models are tested on various samples. A study on entrepreneurship students at the last phase of total early-stage entrepreneurial activity (TEA) indicated ESE is the most significant factor for entrepreneurial intention and behavior [11].

Role models have a significant and positive impact on entrepreneurial intention. While gender moderated the effects, self-efficacy mediated the relations. Role models had a stronger influence on women's self-efficacy, which, in turn, influenced their career intention [71]. In the case of women also, the ESE is a strong and useful mediator of the effect of informal institutional support on venture performance [72].

The relationship between dynamic capability and self-efficacy was explored in an in-depth qualitative study that indicated that perceived self-efficacy could influence dynamic capability. The perceived self-efficacy was proposed as an essential component of dynamic capabilities [73].

Self-efficacy is also investigated as an antecedent of entrepreneurial intention (EI). Of the personal and contextual determinants, a situational approach indicated that among the factors of internal locus of control (LoC), self-regulation, and support of family and friends, self-efficacy completely mediated the relationship between internal LoC, self-regulation, and EI but partially mediated the relationship between support from family and friends and EI. Gender-related differences were also observed in the study [74].

In the case of nascent entrepreneurs also, the confidence in one's ability to execute tasks required in entrepreneurship was found to be a robust predictor of start-up, whereas the outcome expectancies played a marginal role [75]. Overconfidence of entrepreneurs is another aspect studied in the self-efficacy context. It was found that business ties promoted entrepreneurs' overconfidence, which encouraged them to use the effectuation method rather than causation in decision-making. Institutional ties influenced entrepreneurs' illusion of control positively. It has been argued that the managerial and institutional ties influence the effectuation and causation behavioral modes where cognitive biases act as a link [37].

Researchers defined resilience as a psychological process, a malleable competency indicating a continued effort irrespective of the outcome of previous attempts [76]. This competency is appropriate to entrepreneurial situations of uncertainty, for example, in the exploitation of opportunity or reentry after a business failure. Researchers identified resilience as a mediator in the relationship between self-efficacy, attitudes, social norms, and entrepreneurial intention. In another research, resilience was defined as persistence in the face of danger in the context of war and terror facing the region. It was found that resilience and self-efficacy reinforce each other for entrepreneurial intention [77].

The relationship between emotional intelligence and ESE has been studied. ESE indicated a gender difference, but emotional intelligence did not differ between genders. ESE and emotional intelligence varied among different grades of education. A significant positive correlation was observed between ESE and emotional intelligence [78].

4.2.5 Leadership Skills

Broadly, the function of the leadership is to get associates to focus on problems that leaders consider important. The skill is exhibited in the interpersonal, decision-making, and organizational context.

Leadership skills, indicating the ability to bridge cognitive social capital, were found to have a strong and significant association with entrepreneurial intentions. It mediated the relationships between the core TPB constructs [17]. However, the research reported mixed results for the role of bonding social capital with the social acceptability of entrepreneurship and family background. The ability of entrepreneurial leadership to infuse creativity among the team members has been studied. It was found that the creative self-efficacy of individuals and teams mediated the relationships between entrepreneurial leadership and the creativity of individual and team. The ability of entrepreneurial leadership was found to influence employee and team creativity, mediated by both employee creative self-efficacy and team creative efficacy. Authors argued that employees and teams led by entrepreneurial leaders are likely to produce creative outcomes [79]. A similar study indicated that entrepreneurial leadership influences the employees' innovative work behavior where the firm's innovative environment mediates the relationship. The relationship between entrepreneurial leadership and employees' innovative behavior was positively moderated by ESE [80].

4.2.6 Values and Ethics

Values are fundamental beliefs responsible for human attitudes or actions. The purposeful process of entrepreneurship is believed to be driven by values. Entrepreneurship is a self-determined human action driven by a specific set of values to make decisions about how to behave in situations. Entrepreneurship was argued to be a self-determined behavior by an individual to express and satisfy different fundamental needs based on four specific values such as independence, creativity, ambition, and daring. These values were argued to be consistent with self-determinism, self-efficacy, and identity, associated with entrepreneurship [81].

Cognitive and stakeholder theories have been used to examine how entrepreneurs decide under uncertainty. The ethical judgment of entrepreneurs is generally lower than those of non-entrepreneurs in moderately complex and novel decision settings where cognitive biases influence the ethical judgment of entrepreneurs [30]. For example, some entrepreneurs decide to harm the natural environment in violation of their values. In such a scenario, they cognitively dis-engage from their pro-environmental values to assess opportunities harming the environment as highly attractive. This aspect is more pronounced when the ESE is high and the industry munificence is low [31].

4.2.7 Growth Mindset Interventions and ESE

Venture growth relates to the survival and sustainability of a new firm. One of the emerging applications of the growth mindset interventions was used in a study. A research reported that the intervention improved the ESE of students and task persistency [82]. The firm growth is believed to have some universal and culture-specific dimensions as antecedents. The understanding of which will help firm survival and sustenance. The resource-based view and SCT were used to theorize that the firm growth is shaped by etic as well as emic forms of entrepreneurial readiness comprising of entrepreneurial capability and willingness [83].

4.2.8 Emotions and Well-being

Emotions are indicated as a strong feeling caused by circumstances, mood, or relationships. The entrepreneurial process with uncertainty, innovation, failure, achievements, and other aspects, therefore, is expected to create strong emotions during the process.

The SCT proposes emotions as important sources of information for people to judge their own efficacy to accomplish specific tasks. An experimental research studied the emotions involved in career starters' anticipatory emotions associated with envisaged business start-up scenarios. They found that independent self-construal was associated with lower intensity of facial expressions of fear; interdependence was associated with higher intensity of anger and disgust. There were gender variations in the process. Authors also indicated the cultural orientation of emotions in the entrepreneurial process [84].

Similarly, theories of grief are used with SCT to identify grief regulation and normalization for coping with negative emotions arising out of project failure in corporate entrepreneurs. Researchers proposed that the higher level of self-efficacy for recovering from grief, social support offered by the organizational environment, and regulating rather than eliminating grief via normalization processes are likely to explain superior learning and motivational outcomes [32].

Persistent and continual seeking of opportunity-laden information satisfies the internal motivators of need for achievement and a competitive urge. The entrepreneurial mind is attracted and stimulated by the elements of excitement and fun from the business challenges matching the skills, knowledge, and abilities. Benefits and potentially positive outcomes are amplified in the calculation process, while potential complications are regarded as challenges to be overcome rather than obstacles to be avoided [27].

During the new business development stage, accessibility of resources leads to entrepreneurial well-being through an ESE mechanism, and an increased entrepreneurial well-being triggers increased persistence with start-up activities [85].

4.3 Family Business

The family business scenario presents substantial contextual differences. The social cognitive approach is also used to understand the family business succession scenario and the role of cognitive complexity in the process [86]. One of the key issues is the success and firm continuity, which depends on the involvement of the next generation.

Researches have indicated that there is an indirect influence of perceived parental support and psychological control on next-generation engagement in family firms through the mediating variables of self-efficacy and commitment to the family business [87].

Similarly, the shaping of the entrepreneurial intention of the next generation and the influence of the family have remained inconsistent. Authors took a novel approach by switching the focus from mere presence to the content of exposure from the family business. They examined how the intention is shaped by the learning experiences and parental rewards involved in the family business. The rewards from parents on entrepreneurial activities were found to influence entrepreneurial intentions, and it was partially mediated by ESE. Interestingly, the positive impact of rewards was weakened if the next generation is directly involved in the business but it strengthens ESE [25].

The interplay of individual factors like passion and self-efficacy with family role models and education in shaping the entrepreneurial intention indicated that self-efficacy plays a significant mediating role [88].

4.4 Intrapreneurship

Intrapreneurship presents a distinct entrepreneurial context and process. In this case, employees act as an entrepreneur within an organization. In some cases, organizations incentivize such efforts. Authors have drawn on career construction theory to understand the antecedents of intrapreneurship. An examination indicated that career adaptability mediated the relations between personality traits and intrapreneurship. The relation between intrapreneurship and both openness and conscientiousness from the Big Five personality dimensions was completely medicated by it. However, career adaptability did not mediate the relationship between intrapreneurship and extraversion, but the direct effects were significant [26].

4.5 Processes

Entrepreneurship involves various decision-making processes, and this section discusses aspects identified in the selected literature.

4.5.1 Entrepreneurial Opportunity

Of late, researchers have tried to identify if social media platforms give the scope for entrepreneurial opportunity identification. Researchers have argued that factors like prior knowledge, alertness, and social media can influence the discovery and the creation of entrepreneurial opportunity, the social media being the moderator [14].

Emotions influence opportunity evaluation and exploitation decisions. Affect-as-information theory indicates that emotions are used as a source of information, and affective processing principles are used for opportunity evaluation and processing. The association between opportunity characteristics and exploitation is mediated by evaluation, whereas both evaluation and emotions influence the exploitation decision. Research found specific emotions such as fear, joy, and anger influenced evaluation as well as exploitation [28].

Researchers also have built models based on the social cognitive career theory and institutional theory and included entrepreneurial cognition factors, contextual factors, entrepreneurial opportunity recognition, and outcome expectations. These models indicated that entrepreneurial self-efficacy as a cognition factor and outcome expectations positively influence opportunity recognition. The self-efficacy and opportunity recognition was moderated by the regulatory and normative contextual factors. However, the institutional factors did not influence the relationship between outcome expectations and opportunity recognition [23].

4.5.2 Information-Processing Framework of Entrepreneurship (I-Pe)

Information processing is one of the fundamental aspects of entrepreneurial cognition and decision processes. An information-processing framework of entrepreneurship (I-Pe) was developed to comprehensively map the entrepreneurial opportunity recognition process. The authors incorporated various entrepreneurial needs, attitudes, and motivations, to understand its impact on the assessment of information cues and diagnosis. The framework described how entrepreneurs process opportunity-related information to arrive at an acceptance or rejection decision of business opportunities [27].

4.5.3 Decision-Making and Firm Performance

In addition to the decision process of opportunity recognition, the decisions for firm growth and performance assume significance in the entrepreneurship literature. It is proposed that the entrepreneurial decision-making may be impacted by the configuration of passion, ESE, and risk perception. The configurational effect was tested to understand if entrepreneurs use causal and effectual decision-making. The study confirmed that a combination of these factors rather than individual factors influences their choice of causal or effectual decision logic [36].

Similarly, the influence of ESE on strategic decision making also has been studied to indicate that the entrepreneurs exhibit a stronger ESE when their ventures make decisions by involving other employees that are more comprehensive and that incorporate more current information [89].

The cognitive biases among entrepreneurs in processes such as acquiring information, evaluation, and comparative judgment formation are indicated as a possible cause of high rates of failure of new ventures. Such studies are based on a biased comparative-judgment-formation framework and the egocentrism theory [29]. The combined influence of individual characteristics of the CEO and the firm-level characteristics does influence firm performance at early stages [90].

4.6 Situation

The entrepreneurial ecosystem seems to influence the total entrepreneurial rate and entrepreneurial innovation if the entrepreneurial attention is higher. R&D transfer and market dynamics negatively predict entrepreneurial rate and innovation. Entrepreneurial finance and taxes or regulation policies influence entrepreneurial innovation positively when entrepreneurship attention is high. Similarly, higher entrepreneurship attention positively influences the total entrepreneurial rate within given physical infrastructure, cultural, and social norms [91].

4.6.1 Role of Region or Location

The growth and sustainability of SMEs have been of research interest. It is suggested that the growth and internationalization of SMEs depend on the attitude of the entrepreneur. But the exact mechanism of how attitude influences such internationalization is not understood well. Studies have considered the TPB and other cognitive factors to suggest the suitability of TPB to understand cognitive antecedents of international business activities. Authors argue that the cognitive aspects can be the starting point to understand the decision-making process for internationalization, and the trait approach or resource-based view contributes little to understand this process [20].

Proposing a theory of international entrepreneurship, researchers investigated the role and consequences of entrepreneurs' immigrant background. A resource-based perspective argued immigrant owners to be more successful than their counterparts in exports. However, the cognitive theory perspective indicated such immigrant owners are prone to overconfidence and thereby have lower financial rewards in the risky international markets [22].

Social cognitive theory is used to extend the immigrant entrepreneurship research on network marketing [10]. A research found that the social environment positively influences the motivation and social competence to start network marketing by the immigrant Chinese entrepreneurs. However, the entrepreneurial action was not found to be influenced by motivation or social competence [10].

Study of internationalization or exporting behavior of entrepreneurs from disadvantaged contexts such as regions facing an economic crisis or developing economies indicates that emotional attributes play a more significant role compared to human capital-related factors. Specifically, the emotional competencies related to social aspects are more important than the personal aspects [92]. In transition economies such as Kazakhstan and Kyrgyzstan, the examination of individual characteristics, environment, and self-efficacy on performance indicated self-efficacy to have a direct and mediating impact on firm performance [93].

The influence of the country-level institutional context on ESE, alertness to opportunity, fear of failure, and opportunity entrepreneurship presents an ambiguous understanding. A study suggested that ESE and alertness encourage opportunity-based entrepreneurship while fear of failure discourages it; the strength of these relationships depends on the institutional context. In countries with higher economic freedom, ESE and alertness can lead to opportunity-driven ventures. Economic freedom helps in mobilizing individuals' social and cognitive resources for high-growth ventures [94]. A contrary finding was reported indicating that the formal institutional support showed no correlation with ESE [72].

4.6.2 Sector

The sector of the chosen business comes with externalities peculiar to it. Tourism entrepreneurship is one of the sectors where authors argued that the nature of personal and environmental factors remains less studied. The study of entrepreneurial motivation, from a dynamic perspective, indicated that the personal factors as cognitive beliefs, intrinsic needs, and demographic factors are important, while changes in the tourism market, changes within the industry, and changes in setting or location are the environmental facilitators [95].

4.7 Dataset, Methods, and Analyses in the Selected Studies

This section discusses the datasets and analyzes methods used by various researchers in their studies. Mostly, the researchers have used collected dataset specific to the research objectives. In few cases researchers have used Global Entrepreneurship Monitor (GEM) surveys, Economic Freedom of the World (EFW) index [94], and OECD data sources as well [91].

Survey methodology has been the predominant mode of data collection [16]. The qualitative research design involved in-depth qualitative study [73], situational approach [74], exploratory analyses [89], and case study [27] in different researches. The fuzzy set-related techniques such as fuzzy set qualitative comparative analysis (fsQCA) [90] and the non-linear neural fuzzy set network model [60] were also used by researchers.

Experimental research design studies such as randomized controlled trial [82], conjoint experiment coupled with hierarchical linear modeling to test a multi-level opportunity recognition model [12], experimental [33], and experimental vignette methodology (EVM) were used [85].

We found various regression techniques used by the researchers for data analyses. The hierarchical linear modeling analysis (HLM) [79], hierarchical regression model [15], seemingly unrelated regression (SUR) [91], and structural equation model [10, 13, 72] were used.

A systematic study of participant's views was assessed in Q methodology [96], and a longitudinal study was also used [91]. The Bayesian path analysis [84] and the confirmatory factor analysis also were found to have been used by the researchers [61].

5 Additional Literature on Cognition Research

This section presents other literature identified through subsequent literature review from the search of specific journals and references identified from the initial search of the literature.

"The Distinctive and Inclusive Domain of Entrepreneurial Cognition Research [97]" and "The Central Question in Entrepreneurial Cognition Research 2007 [98]" are two seminal papers on cognition research trying to clarify the broad outline, approaches, and suggestions to researchers.

Authors have tried to delineate the boundary of entrepreneurship cognition research. Concerns also have been raised that entrepreneurship cognition research should not be a net importer of ideas from the parent discipline of cognitive psychology; instead, it should be able to generate interesting research questions and contribute back. Researchers from other domains should also use the ideas developed in the entrepreneurship domain. They also indicate broad questions on entrepreneurial cognition researches. Why some choose to become entrepreneurs, why few sense the opportunity, how they make decisions for competitive advantage, and how the cognitive aspects can be measured in a social setting are some of the important questions on entrepreneurial cognition research. Creative cognition, cognitive heuristics, mental simulations, counterfactual thinking, opportunity identification, the process of discovery, evaluation, and exploitation, the role of cognitive map, and measurement models are some of the indicated domains for research. The authors have distinguished it from the effectuation process indicated by Sarasvathy. Given the scope of entrepreneurship research, these processes can be investigated at the individual (i.e., entrepreneur), firm, and community level. A large number of entrepreneurial failures also merit attention of researchers. The advantage of such investigation in entrepreneurship is argued to be without much "noise" and confounding effect. They also have indicated that instead of recall of past situations, the actual thinking of entrepreneurs can be investigated for better results. The article suggests the article "Cognitive Approaches to New Venture Creation" as a starting point [99].

Mitchell et al. [98] further delineated on entrepreneurship and cognition research in their article "The Central Question in Entrepreneurial Cognition Research 2007." The basic question "How do entrepreneurs think?" has been exploded to different questions in the ambit of entrepreneurial cognition research. At the individual level, questions like why few chose to become entrepreneurs, a few recognize opportunities, and still, a few become more successful compared to others are some of the apt questions for cognition research assuming differences in mental processes. Entrepreneurial thinking is assumed to be distinctive. Since the entrepreneurship processes take place in a socioeconomic setting, the role of social cognition also assumes significance. Some of the theories such as entrepreneurial heuristics theory, entrepreneurial alertness theory, entrepreneurial expertise or script theory, and effectuation theory are more relevant bases for further research.

The difference between causation and effectuation approach, deep-seated belief, belief structures, and changes to belief influencing entrepreneurial thinking, while moving toward a professional mindset are some of the interesting questions for research. Under the person category of cognition research, how entrepreneurs learn and develop their cognitive structures, the superiority of any specific method of learning, the goal of economic and noneconomic value and differing thoughts on context, how team interactions change entrepreneurial cognition, the role of affective experiences on entrepreneurial cognition are some of the proposed cognition researches.

Under the context or situation category, the role of heuristics on entrepreneurial decision-making, the role of effectuation vs logical goal-directed behavior, and the role of governance, incentives, and environments on the choice of specific cognitive logic for entrepreneurial processes remain some of the intriguing aspects of cognition research.

Under the cognition category, the role of metacognition, metamemory, metacomprehension, and self-regulation in shaping and subsequent changing of the mental models remains broad areas of research. Under the motivation category, personal motives in for-profit and not-for-profit situations, the interaction of various motives, and the relative importance of various motives, in various process of entrepreneurship, need investigation.

6 Limitations

We considered only two databases to search for articles and subsequently selected article through a purposeful search; the hybrid model has advantages and disadvantages inherent in this study. Secondly, the search criteria included entrepreneurship and cognition, which plausibly excluded cognition research in other areas with possible influence in entrepreneurship.

7 Future Research Direction

One of the sharpest criticisms of entrepreneurial cognition research indicated that mistakes of trait research have not been getting addressed by the researchers [3]. It is suggested that variables external to the entrepreneur have to be used to explain the individual behavior, though a few have started using contextual variables. Secondly, very few cognitive variables are being researched upon for a few specific entrepreneurial phases causing the lack of comprehensive or coherent understanding. Thirdly, the definition of an entrepreneur is restrictive since only the business creators are studied [3].

One of the major challenges is to exclude the individual focus in entrepreneurship research. Focus on the process, context, and outcomes, advocated long back, remains relevant for entrepreneurship research. Subsequent researchers also emphasized that the theoretical shift from individual to consequences of actions, peculiar use of knowledge, network, and resources by the entrepreneur, an improved taxonomy of forces, and different levels of analysis are expected to show deeper insight into entrepreneurship. Since it is a recursive and continuous process, the integration to research design remains a challenge, but a requirement [100]. Since many individuals think of entrepreneurship options, few try, and still fewer become successful, entrepreneurship can be thought of as an instantiation of specific cognitive processes, contexts, and resources with specific outcomes.

The review finds that most researches and models have based their studies on the SCT. However, scientists argue that social cognition may not have a physical basis as in having a special neural system. The same neural systems perform similar functions while processing information from social as well as non-social domains [5]. This aspect may indicate a fundamental mismatch of the physical basis and conceptual models of cognitive theory. However, the findings and arguments are inconclusive; contrary evidences are available and computational neuroscience models have been able to mimic socio-cognitive processes [101–104].

Cognitive variables considered in entrepreneurship research are most likely to be flow variables rather than stock variables. The dynamism of interaction, over a period, may change the state of the variable introducing a classical uncertainty of objective observation. Various studies included in this research regarding risk taking, uncertainty, resilience, and persistence belong to this aspect and require further comprehensive understanding. Similarly, the environmental uncertainty, resilience, and career choice [76], observing resource gain spiral expected in the entrepreneurial process [13], and testing of ESE dimensions and its subsequent impact [67] require longitudinal perspectives. Antecedents of self-efficacy, its change over a period, associated negative effects, alternate theoretical approaches, collective self-efficacy, and measurements remain thought-provoking areas for future research [105].

Answering the big question of "How do entrepreneurs think?" requires what information was available, what assumptions were made, what mental model they used, and how it interacted with available information to transform to be of entrepreneurial value. More significantly, at a higher level of cognition, why did they chose the specific action out of several alternatives they generated? Entrepreneurship cognition research has the possibility of answering such questions more convincingly.

Index: Journal included in the study in alphabetical order

Journal	#	Publisher
Academy of Entrepreneurship Journal	1	Abacademies
Academy of Management Journal	1	AOM
Asia Pacific Business Review	1	Taylor and Francis
Asia Pacific Journal of Innovation and Entrepreneurship	1	Emerald
Australasian Marketing Journal	1	Elsevier
Career Development International	2	Emerald
Current Psychology	1	Springer
Entrepreneurship Research Journal	1	De Gruyter
Entrepreneurship: Theory and Practice	8	Sage
European Journal of Innovation Management	1	Emerald
European Journal of Training and Development	1	Emerald
Expert Systems with Applications	1	Elsevier
Frontiers in Psychology	4	Frontiers
Gender and Education	1	Taylor and Francis
International Business Review	1	Elsevier
International Entrepreneurship and Management Journal	3	Springer
International Journal of Electronic Finance	1	Inderscience
International Journal of Entrepreneurial Behaviour and Research	5	Emerald
International Journal of Entrepreneurial Venturing	1	Inderscience
International Journal of Environmental Research & Public Health	1	MDPI
International Journal of Gender and Entrepreneurship	1	Emerald
International Journal of Hospitality Management	1	Elsevier
International Journal of Management Education	1	Elsevier
International Journal of Manpower	1	Emerald
International Journal of Technology and Design Education	1	Springer
International Journal of Technology Management	1	Inderscience
International Small Business Journal: Researching Entrepreneurship	3	Sage
Journal of Business Venturing	7	Elsevier
Journal of Applied Social Psychology	2	Wiley
Journal of Business and Psychology	1	Springer
Journal of Business Research	3	Elsevier
Journal of Career Development	1	Sage
Journal of Developmental Entrepreneurship	1	World Scientific
Journal of Enterprising Culture	2	World Scientific

Journal	#	Publisher
Journal of Entrepreneurial Behaviour and Research	1	Emerald
Journal of Entrepreneurship	2	Sage
Journal of Global Entrepreneurship Research	1	Springer
Journal of International Business Studies	1	Springer
Journal of International Entrepreneurship	2	Springer
Journal of International Management	1	Elsevier
Journal of Occupational and Organizational Psychology	1	Wiley
Journal of Product Innovation Management	1	Wiley
Journal of Small Business and Enterprise Development	2	Emerald
Journal of Small Business Management	1	Wiley
Journal of Vocational Behavior	1	Elsevier
Management Science Letters	1	Growing Science
PLoS ONE	1	PLOS
Psychology Research and Behavior Management	1	Dove Press
Revista Argentina de Clinica Psicologica	1	Revista Argentina
Small Business Economics	2	Springer
Social Behaviour and Personality	1	SBP
Strategic Entrepreneurship Journal	1	Wiley
Studies in Higher Education	1	Taylor and Francis
Sustainability	1	MDPI
Thinking Skills and Creativity	1	Elsevier
Total	88	

References

1. Mitchell, R.K., Busenitz, L., Lant, T., McDougall, P.P., Morse, E.A., Smith, J.B.: Toward a theory of entrepreneurial cognition: rethinking the people side of entrepreneurship research. Entrep. Theory Pract. **27**(2), 93–104 (2002). https://doi.org/10.1111/1540-8520.00001
2. Buchanan, J.M., Di Pierro, A.: Cognition, choice, and entrepreneurship. South. Econ. J. **46**(3), 693–701 (1980)
3. Sánchez, J.C., Carballo, T., Gutiérrez, A.: The entrepreneur from a cognitive approach. Psicothema. **23**(3), 433–438 (2011). www.psicothema.com
4. Neisser, U.: Cognitive sychology. Appleton-Century-Croft (1967). https://scholar.google.com/scholar?hl=en&as_sdt=0,5&q=Neisser,+U.+(1967).+Cognitive+psychology.+New+York:+Appleton-Century-Crofts
5. Beer, J.S., Ochsner, K.N.: Social cognition: a multi level analysis. Brain Res. **1079**(1), 98–105 (2006). https://doi.org/10.1016/j.brainres.2006.01.002
6. Muñoz, P.: A cognitive map of sustainable decision-making in entrepreneurship: a configurational approach. Int. J. Entrep. Behav. Res. **24**(3), 787–813 (2018). https://doi.org/10.1108/IJEBR-03-2017-0110
7. Snyder, H.: Literature review as a research methodology: an overview and guidelines. J. Bus. Res. **104**, 333–339 (2019). https://doi.org/10.1016/j.jbusres.2019.07.039

8. Bacq, S., Ofstein, L.F., Kickul, J.R., Gundry, L.K.: Perceived entrepreneurial munificence and entrepreneurial intentions: a social cognitive perspective. Int.Small Bus. J. Res. Entrep. **35**(5), 639–659 (2017). https://doi.org/10.1177/0266242616658943
9. Caines, V., Earl, J.K., Bordia, P.: Self-employment in later life: how future time perspective and social support influence self-employment interest. Front. Psychol. **10**, 448 (2019). https://doi.org/10.3389/fpsyg.2019.00448
10. Dai, F., Wang, K.Y., Teo, S.T.T.: Chinese immigrants in network marketing business in Western host country context. Int. Bus. Rev. **20**(6), 659–669 (2011). https://doi.org/10.1016/j.ibusrev.2011.02.015
11. Darmanto, S., Yuliari, G.: Mediating role of entrepreneurial self efficacy in developing entrepreneurial behavior of entrepreneur students. Acad. Entrep. J. **24**(1), (2018). https://www.abacademies.org/articles/mediating-role-of-entrepreneurial-self-efficacy-in-developing-entrepreneurial-behavior-of-entrepreneur-students-7055.html
12. Kushev, T., Ahuja, M.K., Carter, R.E.: A social cognitive perspective on opportunity evaluation. J. Entrep. **28**(1), 35–67 (2019). https://doi.org/10.1177/0971355718810570
13. Laguna, M., Razmus, W., Żaliński, A.: Dynamic relationships between personal resources and work engagement in entrepreneurs. J. Occup. Organ. Psychol. **90**(2), 248–269 (2017). https://doi.org/10.1111/joop.12170
14. Park, J.Y., Sung, C.S., Im, I.: Does social media use influence entrepreneurial opportunity? A review of its moderating role. Sustainability (Switzerland) **9**(9), (2017). https://doi.org/10.3390/su9091593
15. Wang, W., Tang, Y., Liu, Y., Zheng, T., Liu, J., Liu, H.: Can sense of opportunity identification efficacy play a mediating role? Relationship between network embeddedness and social entrepreneurial intention of university students. Front. Psychol. **10**, 1342 (2019). https://doi.org/10.3389/fpsyg.2019.01342
16. Li, L., Wu, D.: Entrepreneurial education and students' entrepreneurial intention: does team cooperation matter? J. Glob. Entrepr. Res. **9**(1), 1–13 (2019). https://doi.org/10.1186/s40497-019-0157-3
17. Henley, A., Contreras, F., Espinosa, J.C., Barbosa, D.: Entrepreneurial intentions of Colombian business students: planned behaviour, leadership skills and social capital. Int. J. Entrep. Behav. Res. **23**(6), 1017–1032 (2017). https://doi.org/10.1108/IJEBR-01-2017-0031
18. Karimi, S., Biemans, H.J.A., Lans, T., Chizari, M., Mulder, M.: Effects of role models and gender on students' entrepreneurial intentions. Eur.J. Train. Dev. **38**(8), 694–727 (2014). https://doi.org/10.1108/EJTD-03-2013-0036
19. Roy, R., Akhtar, F., Das, N.: Entrepreneurial intention among science & technology students in India: extending the theory of planned behavior. Int. Entrep. Manag. J. **13**(4), 1013–1041 (2017). https://doi.org/10.1007/s11365-017-0434-y
20. Sommer, L.: Internationalization processes of small- and medium-sized enterprises-a matter of attitude? J. Int. Entrep. **8**(3), 288–317 (2010). https://doi.org/10.1007/s10843-010-0052-z
21. Prajapati, K., Biswas, S.N.: Effect of entrepreneur network and entrepreneur self-efficacy on subjective performance: a study of handicraft and handloom cluster. J. Entrep. **20**(2), 227–247 (2011). https://doi.org/10.1177/097135571102000204
22. Morgan, H.M., Sui, S., Baum, M.: Are SMEs with immigrant owners exceptional exporters? J. Bus. Ventur. **33**(3), 241–260 (2018). https://doi.org/10.1016/j.jbusvent.2017.12.003
23. Camelo-Ordaz, C., Diánez-González, J.P., Franco-Leal, N., Ruiz-Navarro, J.: Recognition of entrepreneurial opportunity using a socio-cognitive approach. Int. Small Bus. J.: Res. Entrep. **38**, 026624262093984 (2020). https://doi.org/10.1177/0266242620939843
24. Santos, S.C., Liguori, E.W.: Entrepreneurial self-efficacy and intentions: outcome expectations as mediator and subjective norms as moderator. Int. J. Entrep. Behav. Res. **26**(3), 400–415 (2019). https://doi.org/10.1108/IJEBR-07-2019-0436

25. Wang, D., Wang, L., Chen, L.: Unlocking the influence of family business exposure on entrepreneurial intentions. Int. Entrep. Manag. J. **14**(4), 951–974 (2018). https://doi.org/10.1007/s11365-017-0475-2
26. Woo, H.R.: Personality traits and intrapreneurship: the mediating effect of career adaptability. Career Dev. Int. **23**(2), 145–162 (2018). https://doi.org/10.1108/CDI-02-2017-0046
27. Pech, R.J., Cameron, A.: An entrepreneurial decision process model describing opportunity recognition. Eur. J. Innov. Manag. **9**(1), 61–78 (2006). https://doi.org/10.1108/14601060610640023
28. Welpe, I.M., Spörrle, M., Grichnik, D., Michl, T., Audretsch, D.B.: Emotions and opportunities: the interplay of opportunity evaluation, fear, joy, and anger as antecedent of entrepreneurial exploitation. Enterp. Theory Pract. **36**(1), 69–96 (2012). https://doi.org/10.1111/j.1540-6520.2011.00481.x
29. Shin, J., Kim, S.K.: The egocentrism of entrepreneurs: bias in comparative judgments. Entrep. Res. J. **9**(1), (2017). https://doi.org/10.1515/erj-2017-0100
30. James, H.S., Ng, D., Klein, P.G.: Complexity, novelty, and ethical judgement by entrepreneurs. Int. J. Entrep. Ventur. **8**(2), 170–195 (2016). https://doi.org/10.1504/IJEV.2016.077624
31. Shepherd, D.A., Patzelt, H., Baron, R.A.: I care about nature, but. Disengaging values in assessing opportunities that cause harm. Acad. Manage. J. **56**(5), 1251–1273 (2013). https://doi.org/10.5465/amj.2011.0776
32. Shepherd, D.A., Covin, J.G., Kuratko, D.F.: Project failure from corporate entrepreneurship: managing the grief process. J. Bus. Ventur. **24**(6), 588–600 (2009). https://doi.org/10.1016/j.jbusvent.2008.01.009
33. Hsu, D.K., Wiklund, J., Cotton, R.D.: Success, failure, and entrepreneurial reentry: an experimental assessment of the veracity of self-efficacy and prospect theory. Enterp. Theory Pract. **41**(1), 19–47 (2017). https://doi.org/10.1111/etap.12166
34. Sweida, G.L., Reichard, R.J.: Gender stereotyping effects on entrepreneurial self-efficacy and high-growth entrepreneurial intention. J. Small Bus. Enterp. Dev. **20**(2), 296–313 (2013). https://doi.org/10.1108/14626001311326743
35. Hechavarria, D.M., Renko, M., Matthews, C.H.: The nascent entrepreneurship hub: goals, entrepreneurial self-efficacy and start-up outcomes. Small Bus. Econ. **39**(3), 685–701 (2012). https://doi.org/10.1007/s11187-011-9355-2
36. Stroe, S., Parida, V., Wincent, J.: Effectuation or causation: an fsQCA analysis of entrepreneurial passion, risk perception, and self-efficacy. J. Bus. Res. **89**, 265–272 (2018). https://doi.org/10.1016/j.jbusres.2018.01.035
37. Zhang, W., Zhao, W., Gao, Y., Xiao, Z.: How do managerial ties influence the effectuation and causation of entrepreneurship in China? The role of entrepreneurs' cognitive bias. Asia Pac. Bus. Rev. (2020). https://doi.org/10.1080/13602381.2020.1765537
38. Crocker, J., Fiske, S.T., Taylor, S.E.: Schematic bases of belief change. In: Attitudinal judgment, pp. 197–226. Springer, New York (1984)
39. Cater, J.J., Young, M.: U.S. Veterans as Emerging Entrepreneurs: Self-Efficacy, Intentions and Challenges. Journal of Developmental Entrepreneurship. **25**(2), 2050012 (2020). https://doi.org/10.1142/S1084946720500120
40. Shahriar, A.Z.M.: Gender differences in entrepreneurial propensity: evidence from matrilineal and patriarchal societies. J. Bus. Ventur. **33**(6), 762–779 (2018). https://doi.org/10.1016/j.jbusvent.2018.04.005
41. Venugopal, V.: Investigating women's intentions for entrepreneurial growth. Int. J. Gend. Entrep. **8**(1), 2–27 (2016). https://doi.org/10.1108/IJGE-11-2014-0043
42. Shahriar, A.Z.M., Shepherd, D.A.: Violence against women and new venture initiation with microcredit: self-efficacy, fear of failure, and disaster experiences. J. Bus. Ventur. **34**(6), (2019). https://doi.org/10.1016/j.jbusvent.2019.06.006
43. Haddoud, M.Y., Onjewu, A.K.E., Nowinski, W., Alammari, K.: Assessing the role of entrepreneurship education in regulating emotions and fostering implementation intention: evidence from Nigerian universities. Stud. High. Educ. (2020). https://doi.org/10.1080/03075079.2020.1758652

44. Barbosa, S.D., Kickul, J., Smith, B.R.: The road less intended: integrating entrepreneurial cognition and risk in entrepreneurship education. J. Enterp. Cult. **16**(04), 411–439 (2008). https://doi.org/10.1142/s0218495808000181
45. Edwards-Schachter, M., García-Granero, A., Sánchez-Barrioluengo, M., Quesada-Pineda, H., Amara, N.: Disentangling competences: interrelationships on creativity, innovation and entrepreneurship. Think. Skills Creat. **16**, 27–39 (2015). https://doi.org/10.1016/j.tsc.2014.11.006
46. Kassean, H., Vanevenhoven, J., Liguori, E., Winkel, D.E.: Entrepreneurship education: a need for reflection, real-world experience and action. Int. J. Entrep. Behav. Res. **21**(5), 690–708 (2015). https://doi.org/10.1108/IJEBR-07-2014-0123
47. Shinnar, R.S., Hsu, D.K., Powell, B.C.: Self-efficacy, entrepreneurial intentions, and gender: assessing the impact of entrepreneurship education longitudinally. Int. J. Manag. Educ. **12**(3), 561–570 (2014). https://doi.org/10.1016/j.ijme.2014.09.005
48. Oo, P.P., Sahaym, A., Juasrikul, S., Lee, S.Y.: The interplay of entrepreneurship education and national cultures in entrepreneurial activity: a social cognitive perspective. J. Int. Entrep. **16**(3), 398–420 (2018). https://doi.org/10.1007/s10843-018-0229-4
49. Brunel, O., Laviolette, E.M., Radu-Lefebvre, M.: Role models and entrepreneurial intention: the moderating effects of experience, locus of control and self-esteem. J. Enterp. Cult. **25**(02), 149–177 (2017). https://doi.org/10.1142/s0218495817500066
50. Baluku, M.M., Matagi, L., Otto, K.: Exploring the link between mentoring and intangible outcomes of entrepreneurship: the mediating role of self-efficacy and moderating effects of gender. Front. Psychol. **11**, 1556 (2020). https://doi.org/10.3389/fpsyg.2020.01556
51. Ratten, V.: Entrepreneurship, e-finance and mobile banking. Int. J. Electron. Finance. **6**(1), 1–12 (2012). https://doi.org/10.1504/IJEF.2012.046590
52. De Carolis, D.M., Saparito, P.: Social capital, cognition, and entrepreneurial opportunities: a theoretical framework. Enterp. Theory Pract. **30**(1), 41–56 (2006). https://doi.org/10.1111/j.1540-6520.2006.00109.x
53. Seet, P.S., Lindsay, N., Kropp, F.: Understanding early-stage firm performance: the explanatory role of individual and firm level factors. Int. J. Manpow. (2020). https://doi.org/10.1108/IJM-02-2019-0112
54. Roy, R., Das, N. Exploring entrepreneurial intention among engineering students in India: a multiple basket approach. Int. J. Technol. Des. Educ. (2020). https://doi.org/10.1007/s10798-020-09596-9
55. Gu, J., Hu, L., Wu, J., Lado, A.A.: Risk propensity, self-regulation, and entrepreneurial intention: empirical evidence from China. Curr. Psychol. **37**(3), 648–660 (2018). https://doi.org/10.1007/s12144-016-9547-7
56. Laguna, M.: Self-efficacy, self-esteem, and entrepreneurship among the unemployed. J. Appl. Soc. Psychol. **43**(2), 253–262 (2013). https://doi.org/10.1111/j.1559-1816.2012.00994.x
57. Biraglia, A., Kadile, V.: The role of entrepreneurial passion and creativity in developing entrepreneurial intentions: insights from American homebrewers. J. Small Bus. Manag. **55**(1), 170–188 (2017). https://doi.org/10.1111/jsbm.12242
58. Yar, D.H., Wennberg, W., Berglund, H.: Creativity in entrepreneurship education. J. Small Bus. Enterp. Dev. **15**(2), 304–320 (2008). https://doi.org/10.1108/14626000810871691
59. Darmanto, S., Pujiarti, E.S.: Developing student's social entrepreneurial intention. Manag. Sci. Lett. **10**(5), 1103–1106 (2020). https://doi.org/10.5267/j.msl.2019.10.032
60. Lin, W.B.: A comparative study on the trends of entrepreneurial behaviors of enterprises in different strategies: application of the social cognition theory. Expert Syst. Appl. **31**(2), 207–220 (2006). https://doi.org/10.1016/j.eswa.2005.09.036
61. Zhou, C.: Analysis on innovation ability of talents based on cognitive theories. Revista Argentina de Clinica Psicologica. **29**(1), 506–512 (2020). https://doi.org/10.24205/03276716.2020.68
62. Liu, J., Chen, J., Tao, Y.: Innovation performance in new product development teams in China's technology ventures: the role of behavioral integration dimensions and collective efficacy. J. Prod. Innov. Manag. **32**(1), 29–44 (2015). https://doi.org/10.1111/jpim.12177

63. Ahlin, B., Drnovšek, M., Hisrich, R.D.: Entrepreneurs' creativity and firm innovation: the moderating role of entrepreneurial self-efficacy. Small Bus. Econ. **43**(1), 101–117 (2014). https://doi.org/10.1007/s11187-013-9531-7
64. Kraemer, R.C.: "Thinking Technology" in mature industry firms: understanding technology entrepreneurship Mariann Jelinek. Int. J. Technol. Manag. **11**(7–8), 799–813 (1996)
65. Ratten, V.: Cloud computing: a social cognitive perspective of ethics, entrepreneurship, technology marketing, computer self-efficacy and outcome expectancy on behavioural intentions. Australas. Mark. J. **21**(3), 137–146 (2013). https://doi.org/10.1016/j.ausmj.2013.02.008
66. Villasana, M., Alcaraz-Rodríguez, R., Alvarez, M.M.: Examining entrepreneurial attributes of Latin American female university students. Gend. Educ. **28**(1), 148–166 (2016). https://doi.org/10.1080/09540253.2015.1093100
67. Drnovšek, M., Wincent, J., Cardon, M.S.: Entrepreneurial self-efficacy and business start-up: developing a multi-dimensional definition. Int. J. Entrep. Behav. Res. **16**(4), 329–348 (2010). https://doi.org/10.1108/13552551011054516
68. Gielnik, M.M., Uy, M.A., Funken, R., Bischoff, K.M.: Boosting and sustaining passion: a long-term perspective on the effects of entrepreneurship training. J. Bus. Ventur. **32**(3), 334–353 (2017). https://doi.org/10.1016/j.jbusvent.2017.02.003
69. Hu, R., Ye, Y.: Do entrepreneurial alertness and self-efficacy predict Chinese sports major students' entrepreneurial intention? Soc. Behav. Pers. **45**(7), 1187–1196 (2017). https://doi.org/10.2224/sbp.6356
70. Cassar, G., Friedman, H.: Does self-efficacy affect entrepreneurial investment? Strateg. Entrep. J. **3**(3), 241–260 (2009). https://doi.org/10.1002/sej.73
71. BarNir, A., Watson, W.E., Hutchins, H.M.: Mediation and moderated mediation in the relationship among role models, self-efficacy, entrepreneurial career intention, and gender. J. Appl. Soc. Psychol. **41**(2), 270–297 (2011). https://doi.org/10.1111/j.1559-1816.2010.00713.x
72. Kazumi, T., Kawai, N.: Institutional support and women's entrepreneurial self-efficacy. Asia Pac. J. Innov. Entrep. **11**(3), 345–365 (2017). https://doi.org/10.1108/apjie-12-2017-041
73. Kevill, A., Trehan, K., Easterby-Smith, M.: Perceiving 'capability' within dynamic capabilities: the role of owner-manager self-efficacy. Int. Small Bus. J.: Res. Entrep. **35**(8), 883–902 (2017). https://doi.org/10.1177/0266242616688523
74. Molino, M., Dolce, V., Cortese, C.G., Ghislieri, C.: Personality and social support as determinants of entrepreneurial intention. Gender differences in Italy. PLoS One. **13**(6), e0199924 (2018). https://doi.org/10.1371/journal.pone.0199924
75. Townsend, D.M., Busenitz, L.W., Arthurs, J.D.: To start or not to start: outcome and ability expectations in the decision to start a new venture. J. Bus. Ventur. **25**(2), 192–202 (2010). https://doi.org/10.1016/j.jbusvent.2008.05.003
76. Pérez-López, M.C., González-López, M.J., Rodríguez-Ariza, L.: Competencies for entrepreneurship as a career option in a challenging employment environment. Career Dev. Int. **21**(3), 214–229 (2016). https://doi.org/10.1108/CDI-07-2015-0102
77. Bullough, A., Renko, M., Myatt, T.: Danger zone entrepreneurs: the importance of resilience and self-efficacy for entrepreneurial intentions. Enterp. Theory Pract. **38**(3), 473–499 (2014). https://doi.org/10.1111/etap.12006
78. Wen, Y., Chen, H., Pang, L., Gu, X.: The relationship between emotional intelligence and entrepreneurial self-efficacy of Chinese vocational college students. Int. J. Environ. Res. Public Health. **17**(12), 1–18 (2020). https://doi.org/10.3390/ijerph17124511
79. Cai, W., Lysova, E.I., Khapova, S.N., Bossink, B.A.G.: Does entrepreneurial leadership foster creativity among employees and teams? The mediating role of creative efficacy beliefs. J. Bus. Psychol. **34**(2), 203–217 (2019). https://doi.org/10.1007/s10869-018-9536-y
80. Li, C., Makhdoom, H.U.R., Asim, S.: Impact of entrepreneurial leadership on innovative work behavior: examining mediation and moderation mechanisms. Psychol. Res. Behav. Manag. **13**, 105–118 (2020). https://doi.org/10.2147/PRBM.S236876

81. Kirkley, W.W.: Entrepreneurial behaviour: the role of values. Int. J. Entrep. Behav. Res. **22**(3), 290–328 (2016). https://doi.org/10.1108/IJEBR-02-2015-0042
82. Burnette, J.L., Pollack, J.M., Forsyth, R.B., Hoyt, C.L., Babij, A.D., Thomas, F.N., Coy, A.E.: A growth mindset intervention: enhancing students' entrepreneurial self-efficacy and career development. Enterp. Theory Pract. **44**(5), 878–908 (2020). https://doi.org/10.1177/1042258719864293
83. Lau, V.P., Dimitrova, M.N., Shaffer, M.A., Davidkov, T., Yordanova, D.I.: Entrepreneurial readiness and firm growth: an integrated etic and emic approach. J. Int. Manag. **18**(2), 147–159 (2012). https://doi.org/10.1016/j.intman.2012.02.005
84. Zampetakis, L.A., Kafetsios, K., Lerakis, M., Moustakis, V.S.: An emotional experience of entrepreneurship: self-construal, emotion regulation, and expressions to anticipatory emotions. J. Career Dev. **44**(2), 144–158 (2017). https://doi.org/10.1177/0894845316640898
85. Marshall, D.R., Meek, W.R., Swab, R.G., Markin, E.: Access to resources and entrepreneurial well-being: a self-efficacy approach. J. Bus. Res. **120**, 203–212 (2020). https://doi.org/10.1016/j.jbusres.2020.08.015
86. Mitchell, J.R., Hart, T.A., Valcea, S., Townsend, D.M.: Becoming the boss: discretion and postsuccession success in family firms. Enterp. Theory Pract. **33**(6), 1201–1218 (2009). https://doi.org/10.1111/j.1540-6520.2009.00341.x
87. Garcia, P.R.J.M., Sharma, P., De Massis, A., Wright, M., Scholes, L.: Perceived parental behaviors and next-generation engagement in family firms: a social cognitive perspective. Enterp. Theory Pract. **43**(2 Special Issue), 224–243 (2019). https://doi.org/10.1177/1042258718796087
88. Hou, F., Su, Y., Lu, M., Qi, M.: Model of the entrepreneurial intention of university students in the Pearl River Delta of China. Front. Psychol. **10**, 916 (2019). https://doi.org/10.3389/fpsyg.2019.00916
89. Forbes, D.P.: The effects of strategic decision making on entrepreneurial self-efficacy. Enterp. Theory Pract. **29**(5), 599–626 (2005). https://doi.org/10.1111/j.1540-6520.2005.00100.x
90. Palmer, C., Niemand, T., Stöckmann, C., Kraus, S., Kailer, N.: The interplay of entrepreneurial orientation and psychological traits in explaining firm performance. J. Bus. Res. **94**, 183–194 (2019). https://doi.org/10.1016/j.jbusres.2017.10.005
91. Yan, Y., Guan, J.: Entrepreneurial ecosystem, entrepreneurial rate and innovation: the moderating role of internet attention. Int. Entrep. Manag. J. **15**(2), 625–650 (2019). https://doi.org/10.1007/s11365-018-0493-8
92. Quintillán, I., Peña-Legazkue, I.: Emotional intelligence and venture internationalization during economic recession. Int. J. Entrep. Behav. Res. **26**(2), 246–265 (2019). https://doi.org/10.1108/IJEBR-08-2018-0521
93. Luthans, F., Ibrayeva, E.S.: Entrepreneurial self-efficacy in Central Asian transition economies: quantitative and qualitative analyses. J. Int. Bus. Stud. **37**(1), 92–110 (2006). https://doi.org/10.1057/palgrave.jibs.8400173
94. Boudreaux, C.J., Nikolaev, B.N., Klein, P.: Socio-cognitive traits and entrepreneurship: the moderating role of economic institutions. J. Bus. Ventur. **34**(1), 178–196 (2019). https://doi.org/10.1016/j.jbusvent.2018.08.003
95. Wang, S., Hung, K., Huang, W.-J.: Motivations for entrepreneurship in the tourism and hospitality sector: a social cognitive theory perspective. Int. J. Hosp. Manag. **78**, 78–88 (2019). https://doi.org/10.1016/j.ijhm.2018.11.018
96. Valliere, D.: Belief patterns of entrepreneurship: exploring cross-cultural logics. Int. J. Entrepr. Behav. Res. **23**(2), 245–266 (2017). https://doi.org/10.1108/IJEBR-12-2015-0297
97. Mitchell, R.K., Busenitz, L., Lant, T., McDougall, P.P., Morse, E.A., Smith, J.B.: The distinctive and inclusive domain of entrepreneurial cognition research. Entrep. Theory Pract. **28**(6), 505–518 (2004). https://doi.org/10.1111/j.1540-6520.2004.00061.x
98. Mitchell, R.K., Busenitz, L.W., Bird, B., Marie Gaglio, C., McMullen, J.S., Morse, E.A., Smith, J.B.: The central question in entrepreneurial cognition research 2007. Entrep. Theory Pract. **31**(1), 1–27 (2007). https://doi.org/10.1111/j.1540-6520.2007.00161.x

99. Forbes, D.P.: Cognitive approaches to new venture creation. Int. J. Manag. Rev. **1**(4), 415–439 (1999). https://doi.org/10.1111/1468-2370.00021
100. Aldrich, H.E., Martinez, M.A.: Many are called, but few are chosen: an evolutionary perspective for the study of entrepreneurship. In: Entrepreneurship: Concepts, Theory and Perspective, pp. 293–311. Springer, Berlin Heidelberg (2007). https://doi.org/10.1007/978-3-540-48543-8_14
101. Amodio, D.M.: Social cognition 2.0: an interactive memory systems account. Trends Cogn. Sci. **23**(1), 21–33). Elsevier Ltd (2019). https://doi.org/10.1016/j.tics.2018.10.002
102. Hackel, L.M., Amodio, D.M.: Computational neuroscience approaches to social cognition. Curr. Opin. Psychol. **24**, 92–97). Elsevier B.V (2018). https://doi.org/10.1016/j.copsyc.2018.09.001
103. Kar, N., Kar, B.: Social cognition and the participatory planning process. Syst. Res. Behav. Sci. (2002). https://doi.org/10.1002/sres.458
104. Kar, N., Kar, B.: Social cognition and individual effectiveness in interpersonal scenarios: a conceptual review. J. Ment. Health Hum. Behav. **22**(1), 27 (2017). https://doi.org/10.4103/0971-8990.210705
105. Newman, A., Obschonka, M., Schwarz, S., Cohen, M., Nielsen, I.: Entrepreneurial self-efficacy: a systematic review of the literature on its theoretical foundations, measurement, antecedents, and outcomes, and an agenda for future research. J. Vocat. Behav. **110**, 403–419). Academic Press Inc (2019). https://doi.org/10.1016/j.jvb.2018.05.012

Impact of Select Cognitive Distortions on Emotional Stress

Ajeya Jha, Akash Kumar Bhoi, Saibal Kumar Saha, Ankit Singh, Samrat Mukherjee, Bibeth Sharma, and Jayarani

1 Introduction

Stress, as a concept, has invited attention of a large number of studies in psychology and management, as it has been identified as a critical manifestation that can put the psychological and physical health of people at risk [1–4]. There are several approaches to understand and mitigate stress, psychological/cognitive; however cognitive behavioral approach is the most studied approach [5, 6]. A cognitive approach explores the association between the individuals and their environment as it is recognized that how an individual evaluates and perceives the stressor event is of substantial importance [7, 8]. Recent insight in this respect has resulted in a new discipline called the positive psychology which involves developing comprehension about how the individual-stressor interaction can interfere in the health and well-being of people [9–11]. Perceived stress is defined as the extent to which an individual experiences life events such as unpredictable, uncontrollable, or overloading [12].

College students have been an interesting subject for stress-based studies. A number of studies underlined mental health issues in students during their studying years at a university [13, 14]. Vulnerability of young people from 12 to 25 years to

A. Jha (✉) · S. K. Saha · S. Mukherjee · B. Sharma · Jayarani
Department of Management Studies, Sikkim Manipal Institute of Technology, Sikkim Manipal University, Sikkim, Majitar, India
e-mail: ajeya.jha@smit.smu.edu.in; saibal.s@smit.smu.edu.in; samrat.k@smit.smu.edu.in; bibeth.s@smit.smu.edu.in; jaya.r@smit.smu.edu.in

A. K. Bhoi
Department of Electrical and Electronics Engineering, Sikkim Manipal Institute of Technology, Sikkim Manipal University, Sikkim, Majitar, India

A. Singh
Symbiosis Institute of Health Sciences, Pune, India

S. R. Samanta et al. (eds.), *Cognitive Computing for Risk Management*, EAI/Springer Innovations in Communication and Computing,
https://doi.org/10.1007/978-3-030-74517-2_2

psychological health has been recorded [15, 16]. Studies also establish that, compared to youngsters of the same age [17] and, over all, to other age groups [11, 18–20], students suffer greater psychological problems. The developmental and transformative nature of college poses a puzzling and perplexing time for students, but it also provides a generous opportunity for evolution and learning life skills that are essential and helpful in college years and beyond.

Studies indicate that stress in college students is related to depression [21–26], anxiety [27, 28], suicide or attempted suicide [29–31], and drinking problems [32–35].

A cognitive distortion *is an exaggerated or irrational thought pattern involved in the onset or perpetuation of psychopathological states, such as depression and anxiety* [36]. The cognitive model explains that we as individuals are presented with negative, positive, and neutral events, which are interpreted with a chain of thoughts, resulting in distinct feelings and moods by individuals. It is to be noted that feelings are an outcome of an individuals' thoughts and of actual events [37]. It, therefore, implies that events when viewed through a negative interpretations lead to unpleasant moods and thoughts. Negative thoughts are a consequence of cognitive distortions originating from negative core beliefs (schema). More than ten cognitive distortions have been identified by various researchers [38, 39] and include *comparison with others*, *dichotomous thinking*, *negative prediction (fortune telling)*, and *externalizing self-worth*. This is true for all including adolescents and college-going students. Association between distorted cognition and stress has been reported for adolescents [40]. It was found that youngsters having greater cognitive distortions perceived life as much more stressful and were susceptible to greater depressive symptoms, than the ones having lower cognitive distortions. This advances support to *diathesis-stress models*. This model indicates that psychopathology is a function of individual cognitive or biological vulnerabilities (i.e., diatheses) and interacting environmental stressors [41, 42].

A positive association between cognitive distortions, distress, and depression is reported by many more researchers [43, 44]. In contrast to this, an early study [45] reported that it is low life stress that results from distortion than the high life stress. Little confusion, therefore, still prevails. A recent study concludes that individual's overall level of cognitive distortion can predict his or her tendency to perceive life events as stressful [46]. Cognitive distortions also act as mediators between life stress and depression in adolescents [40].

Cognitive distortions and irrational beliefs have also been studied widely in this respect [47–50]. Irrational beliefs have also been found associated with stress and emotional problems [51–54]. A recent study [55] has explored the dynamics of irrational beliefs and emotional disturbances in university students. It also concludes that association exists between irrational beliefs and depression, anxiety, and stress.

This chapter is based on research undertaken to explore the relationship between select cognitive distortions and stress in a college setting. Results may be useful to mitigate the negative consequences of emotional stress born of cognitive distortions.

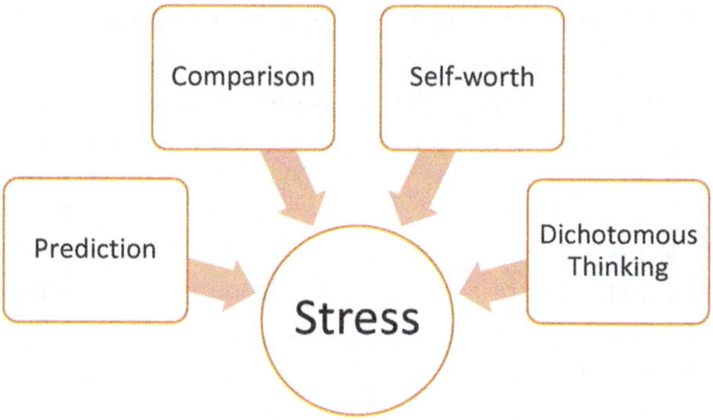

Fig. 1 A proposed model for the study

2 Methodology

This study has been aimed to determine the impact of four select constructs of cognitive distortion, namely, comparison *to others*, *dichotomous thinking*, *prediction* or *fortune telling*, and *externalization of self-worth* on emotional stress of college-going students. The scales for cognitive distortion used for the study have been derived from the inventory proposed by Roberts, M. B. [46]. Considering the constraints it has been shortened, but as the analysis indicates, its validity remains high. The scale for emotional stress, Depression Anxiety Stress Scales (DASS) [56], has been the input with only the stress component being utilized for this study. Components of stress included are *difficult to relax*, *upset easily*, *overreact*, and *impatient*. A proposed model of the study has been given in Fig. 1.

As stress is an outcome of many variables, the study was confined to college students with similar financial background so that focus could remain primarily upon stress arising from cognitive distortions. Both the scales carry values from 0 to 4 with the lowest and highest possible scores being 0 and 40, respectively. The data has been collected from 400 respondents. As 19 questionnaires were filled just partially, the final sample size is 381. Non-response has been relatively low at around 12%. Geographical location on which the survey has been carried out is part of India, commonly known as Northeast India which is a distinct socioeconomic entity.

3 Result and Discussion

Correlation Matrix (Cognitive Distortion Factors (CDF)) To analyze the data, the first step comprised of determining the bivariate correlation that the ten variables under study share. Table 1 reflects the results in this respect.

From the table it may be observed that the two variables that constitute the latent construct *comparison to others*, a component of cognitive distortion, share a correlation of 0.577. For humanities studies, this reflects a moderately strong correlation.

Table 1 Correlation matrix (cognitive distortion factors)

	Comp1	Comp2	Dico1	Dico2	Dico3	Pred1	Pred2	Inper1	Inper2	Inper3
Compare1	1.000	0.577	0.049	0.083	0.140	0.115	0.100	0.081	0.150	0.173
Compare2		1.000	0.090	0.144	0.088	0.117	0.137	0.078	0.096	0.109
Dicoto1			1.000	0.663	0.533	0.165	0.118	0.047	0.052	0.008
Dicoto2				1.000	0.655	0.134	0.148	0.032	0.032	0.016
Dicoto3					1.000	0.067	0.108	0.076	0.111	0.061
Predict1						1.000	0.725	0.042	0.104	0.093
Predict2							1.000	0.110	0.116	0.076
Selfworth1								1.000	0.681	0.670
Selfworth2									1.000	0.774
Selfworth3										1.000

With other variables measuring different latent construct, the correlations range from 0.049 to 0.173, and this implies that the two variables either share no correlation with other variables or at best share weak correlation. The discriminant validity appears to be high.

Three variables that comprise *dichotomous thinking* appear next; between these the correlation coefficients are 0.663 and 0.533. This shows that their correlations are moderately strong, and later when subjected to exploratory factor analysis, they may fall in the same component. Their correlation coefficients with other variables have a range of 0.008–0.165 which indicate no correlation or weak correlation. Again discriminant validity is high.

The third set of two variables measures the latent construct *prediction* or *fortune telling* as it is commonly referred. Table 1 reflects their coefficient correlation which stands at 0.725. For studies pertaining to management, this value reflects high correlation, and again we presume that these will emerge together during factor analysis. It is to be noted that once more the correlation coefficient of these with other variables is 0.042–0.165. The resulting interpretation is a none or low correlation indicating high discriminant validity.

Finally we arrive at three variables that together compose *externalization of self-worth*. These share a correlations stretching from 0.681 to 0.774, implying a moderate or high correlation shared. As in previous cases, we expect these to fall in the same component with factor analysis. Their correlation coefficients with other variables range from 0.008 to 0.116, indicating a nil correlation. As in the earlier case, it reflects high discriminant validity.

Now that an understanding of correlation coefficient has developed, the next step is to subject the data to exploratory factor analysis as this would help in confirming the assumptions stated earlier.

Exploratory Factor Analysis (Cognitive Distortion Factors CDF)) Correlation coefficients determine if the data can be subjected to regression or not. To obtain reliable regression model, it requires confirmation whether the underlying variables comprise latent construct with the available data or not. To perform factor analysis, the first step is to determine sampling adequacy and suitability of the data for

Table 2 KMO and Bartlett's test (CDF)

Kaiser-Meyer-Olkin measure of sampling adequacy		0.654
Bartlett's test of sphericity	Approx. chi-square	1556.673
	df	45
	Sig.	0

structure detection. This is done using KMO and Bartlett's test. Table 2 provides the salient details.

It is to be noted that that KMO yields a value of 0.654. KMO values above 0.6 imply that sampling is adequate and that no corrective measures are needed. As for factor analysis to be functional, certain relationships between variables need to exist; therefore, a significant Bartlett's test of sphericity is essential. In this case it is set at 0.05. Almost zero significance, following a high chi-square value (1556.673), indicates that the hypothesis that there is no significant correlation between variables is rejected and that factor analysis will be beneficial.

An understanding of total variance is needed before reaching factor matrix. In this case Table 3 shows that the four components had eigenvalues above 1 with the first component's eigenvalue being 2.788. It accounts for as much as 27.879% of variance. As it is much below 50%, high discriminant validity is indicated. Together the four components explain 79.7% of variance which is high for this kind of study.

Finally we arrive at the results of factor analysis (Table 4). Rotation method used is Varimex with Kaiser normalization. Earlier extraction method utilized was principal component analysis.

From the matrix it is found that the ten variables fall neatly in four components and with high factor loadings (above 0.8). This is not surprising as validated scale has been used for this study and factor analysis has been undertaken just to confirm suitability of our data. This provides confidence to go ahead with regression analysis to predict stress levels (dependent variable) using the four cognitive distortions.

Exploratory Factor Analysis (Emotional Stress EF) As mentioned, emotional stress is measured by the stress component of Depression Anxiety Stress Scale and is comprised of four factors, namely, *difficult to relax*, *upset easily*, *overreact*, and *impatient*. In order to determine the suitability of data for subjecting it to factor analysis, KMO and Bartlett's tests are the prerequisites and have been shown in Table 5. KMO value of 0.686 and the corresponding chi-square value of 1792 (for Bartlett's test) are sound indicators to go ahead with factor analysis for emotional stress factors.

Having confirmed the adequacy of and suitability of data, the next step involves extracting the data which has been done using principal component analysis (Table 6) . We find four components with eigenvalue more than 1 with the first component having an eigenvalue of 3.369 and explaining as much as 33.695% of variance. Taken together the four components account for as much as 81.246% of variance which is certainly quite healthy. Having satisfied with the prerequisites, the final step has been to rotate the component matrix (using *varimax* in this case).

Table 3 Total variance explained (CDF)

	Initial eigenvalues		
Component	Total	% of variance	Cumulative %
1	2.788	27.879	27.879
2	2.193	21.934	49.813
3	1.592	15.924	65.738
4	1.397	13.973	79.711

Extraction method: principal component analysis

Table 4 Rotated component matrix (CDF)

	Component			
	1	2	3	4
Compare1				0.883
Compare2				0.879
Dicoto1		0.843		
Dicoto2		0.895		
Dicoto3		0.840		
Predict1			0.924	
Predict2			0.919	
Selfworth1	0.870			
Selfworth2	0.909			
Selfworth3	0.904			

Extraction method: principal component analysis
Rotation method: varimax with Kaiser normalization

a. Rotation converged in 4 iterations

Rotated Component Matrix (EF) Rotated component matrix (Table 7) provides a clear insight into the factors obtained. It is to be noted that all the ten variables fall neatly into four components with component 1 presenting the three variables that comprise the latent construct *upset easily*; component 2 carrying the three variables constituting the latent construct *impatient*; component 3 holding the two variables that constitute *difficult to relax*; and the last component displaying two variables that form the latent construct *overreact*.

Regression Analysis Having satisfied the basic requirement of factor analysis to perform regression, the next step involved estimating factor scores of the latent constructs from the values of variables that comprise them. There are several ways to achieve it. For this work, Thomson's method [57] has been followed. This method determines the factor score as function of three terms:

(a) The factor loading matrix
(b) The inverse of the data covariance matrix
(c) The data vector of interest

Table 5 KMO and Bartlett's test (EF)

Kaiser-Meyer-Olkin measure of sampling adequacy		0.686
Bartlett's test of sphericity	Approx. chi-square	1792
	df	45
	Sig.	0

Table 6 Total variance explained (EF)

	Initial eigenvalues		
Component	Total	% of variance	Cumulative %
1	3.369	33.695	33.695
2	1.895	18.953	52.648
3	1.584	15.842	68.491
4	1.276	12.756	81.246
Extraction method: principal component analysis			

Factor scores of four latent constructs, namely, (a) comparison with others, (b) dichotomous thinking, (c) prediction (fortune telling), and (d) externalizing self-worth, have been calculated accordingly.

The next step involved determining the values of R, R^2, and adjusted R^2. From Table 8 these are found to be 0.709, 0.502, and 0.497. R value of 0.709 is fairly high and provides the basis for regression. R^2 value is 0.502, and this implies that the developed regression model will explain 50.2% of variance. This is low but acceptable for humanities studies. Adjusted R^2 is useful to judge goodness of model. It penalizes for adding a variable that does provide any useful contribution to the regression model. Difference between R^2 and adjusted R^2 should be within 0.05 to ensure that no useless variable has been added. In the present case, the difference between the two is 0.005, signifying a good fit for the model. Next important test is the Durbin Watson test to check the presence of autocorrelation. In the present case, the possibility of auto-correlation has been nil; still the test has been done. Durbin Watson yields a value ranging from 1 to 4. A value of 2 indicates zero correlation. Values below 2 imply positive and values above two signify negative correlation. Durbin Watson values between 1.5 and 2.5 are normal. Beyond this range these suggest caution. In this case a value of 2.031 is very healthy. To test the goodness of model, ANOVA is undertaken. Goodness is implied from high value of F. In the existing case, F value at 94.51 is rather assuring. This is confirmed by the corresponding p value which is almost 0, far below 0.05.

Having confirmed the goodness of anticipated regression model, the next step involves developing the regression model. Table 6 provides the same.

It has to be seen that regression model values (Table 9) in this instance are 8.627 (constant), 0.451 (comparing with others), 0.594 (dichotomous thinking), 0.514 (prediction/fortune telling), and 0.609 (externalizing self-worth). Before these are expected, the t value needs to be looked into. A t value below 1.96 (at 95% CL) will render these variables inadequate. We, however, find that in this case all carry values far above the tabulated t value. The last column depicts values for variance inflation

Table 7 Rotated component matrix (EF)

	Component			
	1	2	3	4
Difrelax1			0.917	
Difrelax2			0.925	
Upset1	0.922			
Upset2	0.869			
Upset3	0.916			
Overreact1				0.870
Overreact2				0.880
Impatient1		0.847		
Impatient2		0.866		
Impatient3		0.872		

Extraction method: principal component analysis
Rotation method: varimax with Kaiser normalization
a. Rotation converged in 5 iterations

Table 8 Model summary

Model	R	R square	Adjusted R square	Durbin Watson	*F* value	*p* value
1	0.709	0.502	0.497	2.031	94.51	0

a. Predictors: (constant), Inpertot, Dicitot, Predtot, Comptot
b. Dependent variable: stress

Table 9 Regression model

	Unstandardized coefficients		Standardized coefficients		
	B	Std. error	Beta	*t* value	VIF
(constant)	8.627	0.6		14.374	
Comptot	0.451	0.088	0.191	5.12	1.05
Dicitot	0.594	0.059	0.375	10.092	1.038
Predtot	0.514	0.084	0.229	6.151	1.048
Selfworth	0.609	0.057	0.394	10.654	1.031

factor (VIF) which provide information about extent of variability within a regression model and form a basis for tests of significance. A general approach for interpreting VIF followed commonly states that (a) 1, not correlated; (b) between 1 and 5, moderately correlated; and (c) greater than 5, highly correlated.

Exactly how large a VIF has to be before it causes issues is a subject of debate. What is known is the more VIF increases, the less reliable regression results become. Conservative authors suggest VIF values above 2.5 indicate regression which should be abandoned. In our case there is well below 1.05 and hence regression results may be accepted. The regression model we have is as follows:

$$\text{Stress} = 8.627 + (0.451 * \text{comparing}) + (0.594 * \text{Dichotomous}) + (0.514 * \text{Fortune telling}) + (0.609 * \text{self worth})$$

4 Conclusion

This study was undertaken to understand the impact of four cognitive distortions, namely, *comparison to others*, *dichotomous thinking*, *prediction* or *fortune telling*, and *externalization of self-worth*, on emotional stress of college-going students. From the results and discussion, it may be concluded that there is moderate impact of these factors on stress. Regression results indicate that all the four independent variables (cognitive distortions) have significant impact on emotional stress with *externalization of self-worth* exerting relatively the strongest impact followed by *dichotomous thinking*. Our study appears to confirm the findings of earlier researchers [45, 58]. *Implications* that are obvious from this study are to recognize that college students are extremely vulnerable to cognitive distortions and irrational thoughts. This should be of a great concern for the teachers and college administrators. A neglect in this respect has severe consequences. In a college setting, students are under the custody of their teachers which implies that the teachers are responsible for their well-being. Sufferings of students because of ignorance or lack of effort on part of teaching community are unacceptable by any standards. The teachers therefore must remain alive to the possibility of students' problems due to cognitive distortions and irrational thoughts. They are responsible to device mechanism to identify such students quickly and then reach out to them quickly for help and support. To this extent, teachers must equip them with information, capabilities, and techniques mentioned later in this sub-section and be in a position to provide positive support instead of remaining helpless.

Second important implication is that cognitive distortions and irrational thoughts have negative consequences for students which should concern the custodian adults in their college. The negative consequences *are anxiety, depression, intoxication, and suicide ideation* [59–68]. Additionally cognitive distortions also adversely affect *career choices* [69–72].

The third implication is that teachers must be alive to this reality and they must equip themselves to challenge and diminish the cognitive distortions of their students in order to safeguard them from the negative consequences of stress. Stress management involves three broad techniques – reactive, proactive, and enactive [73]. Reactive techniques bring down stress levels instantly, are easy to learn and practice, but have just a short-term effect. Proactive techniques are built around enhancing the resilience of individuals and are moderate in terms of results, effort, and longevity of effect. Enactive strategies, interestingly, are difficult to learn and require a long time to practice but can eliminate possibility of future stress in certain respects. One such technique is cognitive behavioral therapy (CBT) and has gained credence in this respect and has been found to be useful [74–78]. A variant of cognitive behavioral therapy that has found a lot of favor is rational emotive therapy (RET). In addition to the general approach RET shares with CBT is a strong humanistic philosophy that espouses natural goodness and freedom of human beings. This implies that led by their cognitive distortions and irrational thoughts, individuals begin to doubt their goodness and free existence. A number of researchers find CBT a powerful tool for mitigating cognitive distortions and related stress [79]. These are oral therapies and do not require professional degree [80]. These, therefore, should be made a apart of teachers' training [81] to ensure adequate capabilities are available to the students.

One of the major limitations of this work to be considered when interpreting the conclusions of the study is that it is based on collection of self-reported data. A negative consequence of self-reported data is its accuracy. Respondents may not express truly felt beliefs and at times may be unable to do so. For example, it is known that during survey of this kind, male students are more reluctant to report the experience of stress or are less aware of stress than female students [82, 83].
At the end it may be stated that cognitive distortions and consequential stress are an important issues for college students and society in general and teachers in particular must strive to provide much-needed safeguard to students, specifically the male students, who are more vulnerable in this regard.

References

1. Chen, W.Q., Wong, T.W., Yu, T.S.: Influence of occupational stress on mental health among Chinese off-shore oil workers. Scand. J. Public Health. **37**(7), 766–773 (2009)
2. Ford, M.T., Matthews, R.A., Wooldridge, J.D., Mishra, V., Kakar, U.M., Strahan, S.R.: How do occupational stressor-strain effects vary with time? A review and meta-analysis of the relevance of time lags in longitudinal studies. Work Stress. **28**(1), 9–30 (2014)
3. Mihaila, T.: Perceived stress scale as a predictor of professional behavior and aspects of well-being. Rom. J. Cogn. Behav. Ther. Hypn. **2**(2), 1–14 (2015)
4. Schaufeli, W.B., Taris, T.W.: A critical review of the Job Demands-Resources Model: implications for improving work and health. In: Bauer, G.F., Hämmig, O. (eds.) Bridging Occupational, Organizational and Public Health a Transdisciplinary Approach, pp. 43–68. Springer, Dordrecht (2014)
5. Rothmann, S., Malan, M.: Work-related well-being of South African hospital pharmacists. S. Afr. J. Ind. Psychol. **37**(1), 1–11 (2011)
6. Scheibe, S., Zacher, H.: The role of emotion and emotion regulation in job stress and wellbeing. Res. Occup. Stress Well Being. **11**, 163–193 (2013)
7. Lazarus, R.S., Folkman, S.: Stress: Appraisal and Coping. Springer, New York (1984)
8. Hirschle, A.L.T., Gondim, S.M.G.: Stress and well-being at work: a literature review. Cien. Saude Colet. **25**, 2721–2736 (2020)
9. Häusser, J.A., Mojzisch, A., Niesel, M., Schulz-Hardt, S.: Ten years on: a review of recent research on the Job Demand-Control (-Support) model and psychological well-being. Work Stress. **24**(1), 1–35 (2010)
10. Malik, S., Noreen, S.: Perceived organizational support as a moderator of affective well-being and occupational stress among teachers. Pak. J. Commer. Soc. Sci. **9**(3), 865–874 (2015)

11. Shallcross, A.J., Troy, A., Mauss, I.B.: Change your feelings or leave them be? (or both?): how best to regulate emotions in the face of stress. In: Scott, R., Kosslyn, S. (eds.) Emerging Trends in Social and Behavioral Sciences, pp. 1–24. John Wiley and Sons, Hoboken (2015)
12. Cohen, S., Kamarck, T., Mermelstein, R.: A global measure of perceived stress. J. Health Soc. Behav. **24**, 385–396 (1983)
13. Blanco, C., Okuda, M., Wright, C., Hasin, D.S., Grant, B.F., Liu, S.-M., et al.: Mental health of college students and their non-college-attending peers: results from the National Epidemiologic Study on Alcohol and Related Conditions. Arch. Gen. Psychiatry. **65**, 1429–1143 (2008)
14. Milojevich, H.M., Lukowski, A.F.: Sleep and mental health in undergraduate students with generally healthy sleep habits. PLoS One. **11**, e0156372 (2016)
15. Windfuhr, K., While, D., Hunt, I., Turnbull, P., Lowe, R., Burns, J., et al.: The national confidential inquiry into suicide and homicide by people with mental illness. Suicide in juveniles and adolescents in the United Kingdom. J. Child Psychol. Psychiatry. **49**, 1155–1165 (2008)
16. Thapar, A., Collishaw, S., Pine, D.S., Thapar, A.K.: Depression in adolescence. Lancet. **379**, 1056–1067 (2012)
17. Roberts, R., Golding, J., Towell, T., Weinreb, I.: The effects of economic circumstances on British students' mental and physical health. J. Am. College Health. **48**, 103–109 (1999)
18. Nerdrum, P., Rustøen, T., Rønnestad, M.H.: Student psychological distress: a psychometric study of 1750 Norwegian 1st–year undergraduate students. Scand. J. Educ. Res. **50**, 95–109 (2006)
19. Walsh, J.M., Feeney, C., Hussey, J., Donnellan, C.: Sources of stress and psychological morbidity among undergraduate physiotherapy students. Physiotherapy. **96**, 206–212 (2010)
20. Moreira, J.F.G., Telzer, E.H.: Changes in family cohesion and links to depression during the college transition. J. Adolesc. **43**, 72–82 (2015)
21. Wallace, D.D., Boynton, M.H., Lytle, L.A.: Multilevel analysis exploring the links between stress, depression, and sleep problems among two-year college students. J. Am. Coll. Health. **65**(3), 187–196 (2017)
22. Suwinyattichaiporn, T., Johnson, Z.D.: The impact of family and friends social support on Latino/a first-generation college students' perceived stress, depression, and social isolation. J. Hispanic Higher Educ., 1538192720964922 (2020)
23. Kumar, B., Shah, M.A.A., Kumari, R., Kumar, A., Kumar, J., Tahir, A.: Depression, anxiety, and stress among final-year medical students. Cureus. **11**(3), e4257 (2019)
24. Choi, S.K.: Convergence relationship between life stress and depression in nursing college students. J. Digit. Converg. **16**(5), 257–266 (2018)
25. Killinger, S.L., Flanagan, S., Castine, E., Howard, K.A.: Stress and depression among veterinary medical students. J. Vet. Med. Educ. **44**(1), 3–8 (2017)
26. Gao, W., Ping, S., Liu, X.: Gender differences in depression, anxiety, and stress among college students: a longitudinal study from China. J. Affect. Disord. **263**, 292–300 (2020)
27. Husky, M.M., Kovess-Masfety, V., Swendsen, J.D.: Stress and anxiety among university students in France during Covid-19 mandatory confinement. Compr. Psychiatry. **102**, 152191 (2020)
28. Mateus, P.C.A.P.H., Oliveira, P.B.C.: Brief biofeedback intervention for stress and anxiety: a study with nursing college students. Int. J. Nurs. **4**(1), 7–12 (2017)
29. Zvolensky, M.J., Jardin, C., Garey, L., Robles, Z., Sharp, C.: Acculturative stress and experiential avoidance: relations to depression, suicide, and anxiety symptoms among minority college students. Cogn. Behav. Ther. **45**(6), 501–517 (2016)
30. Lew, B., Huen, J., Yu, P., Yuan, L., Wang, D.F., Ping, F., et al.: Associations between depression, anxiety, stress, hopelessness, subjective well-being, coping styles and suicide in Chinese university students. PLoS One. **14**(7), e0217372 (2019)
31. Kim, H.J.: The effects of college students' stress on drinking, smoking, and health promoting behaviors. Ind. J.Public Health Res. Dev. **10**(5), 1065–1071 (2019)
32. Perrotte, J.K., Baumann, M.R., Knight, C.F.: Traditional gender roles and the stress–alcohol relationship among Latina/o college students. Subst. Use Misuse. **53**(10), 1700–1705 (2018)

33. Luk, J.W., Fairlie, A.M., Lee, C.M.: Daily-level associations between negative mood, perceived stress, and college drinking: do associations differ by sex and fraternity/sorority affiliation? Subst. Use Misuse. **53**(6), 989–997 (2018)
34. Clarke, K.L.: Multivariate relationships of binge watching-drinking-eating with depression, anxiety, and stress in college students. Cogn. Ther. Res. **2**, 189–192 (2019)
35. Kim, H.J., Kim, H.Y.: Emotional labor, job stress and professional quality of life among nurses in long-term care hospital. Korean J. Adult Nurs. **29**(3), 290–301 (2017)
36. Helmond, P., Overbeek, G., Brugman, D., Gibbs, J.C.: A meta-analysis on cognitive distortions and externalizing problem behavior. Crim. Justice Behav. **42**(3), 245–262 (2015)
37. Burns, D.D.: Feeling Good. The New Mood Therapy. Willian Morrow and Company. Inc, New York (1980) (2)
38. Freeman, A.M.: The 10 Dumbest Mistakes Smart People Make and how to Avoid them: Simple and Sure Techniques for Gaining Greater Control of your Life. HarperCollins, New York (1992)
39. Freeman, A., Oster, C.: Cognitive behaviour therapy. In: Herson, M., Bellack, A.S. (eds.) Handbook of Interventions for Adult Disorders, 2nd edn, pp. 108–138. Wiley and Sons, New York (1999) Deal
40. Deal, S.L., Williams, J.E.: Cognitive distortions as mediators between life stress and depression in adolescents. Adolescence. 23(90), 477 (1988)
41. Cicchetti, D., Rogosch, F.A.: Equifinality and multifinality in developmental psychopathology. Dev. Psychopathol. **8**, 597–600 (1996). https://doi.org/10.1017/S0954579400007318
42. Lazarus, R.S.: From psychological stress to the emotions: a history of changing outlooks. Annu. Rev. Psychol. **44**, 1–21 (1993). https://doi.org/10.1146/annurev.psych.44.1.1
43. Smith, T.W., O'Keeffe, J.L., Christensen, A.J.: Cognitive distortion and depression in chronic pain: association with diagnosed disorders. J. Consult. Clin. Psychol. **62**(1), 195 (1994)
44. Lefebvre, M.F.: Cognitive distortion and cognitive errors in depressed psychiatric and low back pain patients. J. Consult. Clin. Psychol. **49**(4), 517 (1981)
45. Hammen, C.L.: Depression, distortion, and life stress in college students. Cogn. Ther. Res. **2**(2), 189–192 (1978)
46. Roberts, M.B.: Inventory of cognitive distortions: validation of a measure of cognitive distortions using a community sample, PCOM Psychology Dissertations. Paper 325 (2015)
47. Tecuta, L., Tomba, E., Lupetti, A., DiGiuseppe, R.: Irrational beliefs, cognitive distortions, and depressive symptomatology in a college-age sample: a mediational analysis. J. Cogn. Psychother. **33**(2), 116–127 (2019)
48. Davies, M.F.: Irrational beliefs and unconditional self acceptance. I. Correlational evidence linking two key features of REBT. J. Ration. Emot. Cogn. Behav. Ther. **24**(2), 113–124 (2006)
49. Stephenson, E., Watson, P.J., Chen, Z.J., Morris, R.J.: Self-compassion, self-esteem, and irrational beliefs. Curr. Psychol. **37**(4), 809–815 (2018)
50. Agahheris, M., Ezzati, K., Dousti, A., Pallooji, A.: Effectiveness of cognitive-behavior training on modifying self and other-deception among females with irrational beliefs. Int. J. Appl. Behav. Sci. **6**(2), 10–18 (2019)
51. Bernard, M.E.: Teacher beliefs and stress. J. Ration. Emot. Cogn. Behav. Ther. **34**(3), 209–224 (2016)
52. Mahfar, M., Xian, K.H., Ghani, F.A., Kosnin, A., Senin, A.A.: Irrational beliefs as mediator in the relationship between activating event and stress in Malaysian fully residential school teachers. Asian Soc. Sci. **14**(10), 2006 (2018)
53. Kaya, D.: Developing the irrational beliefs in mathematics scale (IBIMS): a validity and reliability study. Int. Electron. J. Elem. Educ. **9**(3), 627–644 (2017)
54. Chan, H.W.: Irrational beliefs, depression, anxiety and stress of university students in Hong Kong. HKU Theses Online (HKUTO) (2016)
55. Chan, H.W.Q., Sun, C.F.R.: Irrational beliefs, depression, anxiety, and stress among university students in Hong Kong. J. Am. Coll. Health, 1–15 (2020)

56. Lovibond, S.H., Lovibond, P.F.: Manual for the Depression Anxiety Stress Scales (DASS). Psychology Foundation Monograph (Available from the Psychology Foundation, Room 1005 Mathews Building, University of New South Wales, NSW 2052, Australia) (1993)
57. Thomson, G.H.: Measuring general intelligence by tests which break the g-hierarchy. Nature. **135**(3402), 71 (1935)
58. Dhanalakshmi, D.: Perceived stress, cognitive distortion, sense of coherence and health among college students. Ind. J. Health Wellbeing. **6**(3), 287 (2015)
59. Yüksel, A., Bahadir-Yilmaz, E.: Relationship between depression, anxiety, cognitive distortions, and psychological well-being among nursing students. Perspect. Psychiatr. Care. **55**(4), 690–696 (2019)
60. Kumar, P., Kumar, P., Mishra, S.D.: Suicidal ideation, cognitive distortions, impulsivity and depression among young adult in Patna, Bihar. J. Psychosoc. Res. **15**(1), 101–112 (2020)
61. Fazakas-DeHoog, L.L., Rnic, K., Dozois, D.J.: A cognitive distortions and deficits model of suicide ideation. Eur. J. Psychol. **13**(2), 178 (2017)
62. Wahesh, E., Myers, J.E.: Treatment of Alcohol Abuse among College Students. Annual Review of Addictions and Offender Counseling: Best Practices, p. 110 (2013)
63. Chung, I.W.: Examining suicidal behavior of Asian American female college students: implications for practice. J. Coll. Stud. Psychother. **18**(2), 31–47 (2004)
64. Bonner, R.L., Rich, A.R.: Toward a predictive model of suicidal ideation and behavior: some preliminary data in college students. Suicide Life Threat. Behav. **17**(1), 50–63 (1987)
65. Shupp, M., Brooks, F., Yost, L.: The interface of drug and alcohol addiction, suicide, and surrender: implications for treatment and approaches. Alcohol. Treat. Q., 1–12 (2020)
66. Ibrahim, N., Amit, N., Din, N.C., Ong, H.C.: Gender differences and psychological factors associated with suicidal ideation among youth in Malaysia. Psychol. Res. Behav. Manag. **10**, 129 (2017)
67. Morpeth, J.E.; Perfectionism and Interpersonal Theory of Suicide: Thwarted Belongingness and Perceived Burdensomeness as Mediators of Multidimensional Perfectionism and Suicide Ideation (2018)
68. Schry, A.R., White, S.W.: Understanding the relationship between social anxiety and alcohol use in college students: a meta-analysis. Addict. Behav. **38**(11), 2690–2706 (2013)
69. Adegun, A.O., Aremu, O.: Effectiveness of career development and cognitive reframe therapy on irrational career thoughts among secondary school students in Ogun State, Nigeria. Afr. Symp. **13**(2), 94–103 (2013, December)
70. Salami, S.O.: Discriminant analysis of academic abilities and work values of career choice of high school adolescents in southwestern Nigeria. J. Pers. Stud. Group Behav. **25**, 49–73 (2005)
71. Sampson Jr., J.P., Reardon, R.C., Peterson, G.W., Lenz, J.G.: Career Counseling and Services: A Cognitive Information Processing Approach. Brooks/Cole, Pacific Grove (2004)
72. Saunders, D.E., Peterson, G.W., Sampson, J.P., Reardon, R.C.: The contribution of depression and dysfunctional career thinking to career indecision. J. Vocat. Behav. **56**, 288–298 (2000)
73. Everly, G.S., Lating, J.M.: Resilience: the final frontier. In: A clinical guide to the treatment of the human stress response, pp. 143–154. Springer, New York (2013)
74. Pedrelli, P., Nyer, M., Yeung, A., Zulauf, C., Wilens, T.: College students: mental health problems and treatment considerations. Acad. Psychiatry. **39**(5), 503–511 (2015)
75. Neenan, M.: Developing Resilience: A Cognitive-Behavioural Approach. Taylor & Francis, ISBN 9780415792912. Published July 4, 2017 by Routledge, 200 Pages (2017)
76. Mahmoud, J.S.R., Staten, R.T., Hall, L.A., Lennie, T.A.: The relationship among young adult college students' depression, anxiety, stress, demographics, life satisfaction, and coping styles. Issues Ment. Health Nurs. **33**(3), 149–156 (2012)
77. Church, D., De Asis, M.A., Brooks, A.J.: Brief group intervention using emotional freedom techniques for depression in college students: a randomized controlled trial. Depress. Res. Treat. **2012**, 257172 (2012)

78. Fitzpatrick, K.K., Darcy, A., Vierhile, M.: Delivering cognitive behavior therapy to young adults with symptoms of depression and anxiety using a fully automated conversational agent (Woebot): a randomized controlled trial. JMIR Mental Health. **4**(2), e19 (2017)
79. Xu, L., Liu, H.: Effects of Rational Emotive Behavior Therapy (REBT) intervention program on mental health in female college students. NeuroQuantology. **15**(4), 156–161 (2017)
80. Samuel C. Ugwoke, , Chiedu Eseadi, , Chima C. Igbokwe, Gertrude T.U. Chiaha, Okechukwu O. Nwaubani, , Chibueze Tobias Orji, Leonard T. Ugwuanyi, Ifeoma S. Chukwuma, Edith C. Edikpa, Vera N. Ogakwu, Eucharia A. Onu, Patricia Agu, P Ujunwa A. Nwobi, Faith Omeke, Francisca C. Okeke, Rita N. Ezema, and Lawretta I. Abugu (2017). Effects of a rational-emotive health education intervention on stress management and irrational beliefs among technical college teachers in Southeast Nigeria. Medicine, 96(31):e7658
81. Nucci, C.: The rational teacher: rational emotive behavior therapy in teacher education. J. Ration. Emot. Cogn. Behav. Ther. **20**(1), 15–32 (2002)
82. Davies, J., McCrae, B.P., Frank, J., Dochnahl, A., Pickering, T., Harrison, B., et al.: Identifying male college students' perceived health needs, barriers to seeking help, and recommendations to help men adopt healthier lifestyles. J. Am. Coll. Health. **48**(6), 259–267 (2000)
83. Brougham, R.R., Zail, C.M., Mendoza, C.M., Miller, J.R.: Stress, sex differences, and coping strategies among college students. Curr. Psychol. **28**(2), 85–97 (2009)

An Empirical Note on Influence of Cognitive Barriers on Drug Adherence

Saibal Kumar Saha, Ajeya Jha, Akash Kumar Bhoi, Ankit Singh, Jitendra Kumar, and Jaya Rani

1 Introduction

As patients do not adhere to the prescribed medication regime, it is an important health challenge. Although there is enough evidence that proper medication adherence leads to improved therapeutic benefits, many patients do not adhere to their regime [1]. Studies report that for common chronic diseases like hypertension and diabetes, the adherence rate is generally between 50% and 65% [2, 3]. Failure to adhere to the medication regime increases the risk of not getting the intended benefits subsequently leading to negative sequelae [4]. Medication non-adherence is also associated with increased healthcare costs [5], increased cost of care, and deleterious consequences for workforce productivity and overall public health [6]. It is estimated that for every additional dollar spent on proper medication, costs related to medical would reduce by $3.98 for hypertension patients, $5.10 for patients suffering from high cholesterol and $7 for diabetic patients [7]. More than 30% of medicine-related hospital admissions have been attributed to medication non-adherence [8].

For treatment of diabetic patients, the process is complex, and it involves following a proper diet regime, regular exercise, regular monitoring of blood glucose level, insulin injections and timely adjustment of the regime for exercise, stress and

S. K. Saha · A. Jha (✉) · J. Kumar · J. Rani
Department of Management Studies, Sikkim Manipal Institute of Technology, Sikkim Manipal University, Sikkim, India
e-mail: jitendra.k@smit.smu.edu.in; jaya.r@smit.smu.edu.in

A. K. Bhoi
Department of Electrical and Electronics Engineering, Sikkim Manipal Institute of Technology, Sikkim Manipal University, Sikkim, Majitar, India

A. Singh
Symbiosis Institute of Health Sciences, Pune, India

S. R. Samanta et al. (eds.), *Cognitive Computing for Risk Management*, EAI/Springer Innovations in Communication and Computing,
https://doi.org/10.1007/978-3-030-74517-2_3

infections as per the situation and severity of the disease [9]. Due to challenges in the methodology, there is paucity in consistent findings with respect to relationship between metabolic control and adherence [10]. According to cognitive behavioural theory, patients respond to cognitive exemplification of a condition and then the condition itself [11]. Evidence suggest that there is mediating effect of cognitive process in the relationship between stress, adherence, and metabolic control [12]. This research work was undertaken with a view to determine if cognitive barriers (denial, refusal and mistrust) affect drug adherence in diabetic people.

2 Literature Review

Medication adherence is defined as "the extent to which a patient act in accordance with the prescribed interval, and dose of a dosing regimen" [13]. Medication persistence is defined as "the duration of time from initiation to discontinuation of therapy" [13]. According to [14], medication adherence can be assessed by direct and indirect methods. In the direct method, the therapy is directly observed and is considered more robust. In the indirect method, the methods of questionnaires, prescription refill rates, pill counts, electronic monitors and patient diary monitoring are used. Self-reported measures are associated with bias of inaccurate patient recall, social desirability and over-estimation of adherence rates. Pill counts are easily monitored with electronic systems and are easy to perform. It is generally associated with randomized clinical trials [15]. This method does not take into account the timing of the medication. Hence, all the measures have their advantages and disadvantages. Published literature suggests that there is no general consensus for best practice/"gold standard" to assess medication adherence [16]. For the elderly patients, it is suggested that patient or caregiver interview using open-ended questions which are non-judgmental nonthreatening should be asked [17].

The World Health Organization has categorized the reasons for medication non-adherence into five factors: "patient, condition, therapy, socioeconomic, and health system–related factors" [18]. Physical and cognitive impairments, psychological, behavioural, age and race are associated with the factor "patient" [19]. Examples of the factor "condition" are asymptotic nature of chronic disease and mental health condition of patients [20]. Dose complexity and adverse side effects are associated with the factor "therapy" [21]. Rate of literacy, costs of medication and social support are associated with the factor "socioeconomic" [22]. The factor "health system" is associated with communication, patient-physician/provider relationship and access to healthcare facilities [23].

Studies report that for youth patients suffering from type 1 diabetes, the risk of consequences in later life (neuropathy, cardiovascular disease, renal failure) decreases if their blood glucose level is controlled in the early stages. However, there is no proper consensus between relationship with life stress and metabolic controls [24]. Research proposed that by physiological mechanisms glycaemic control could be directly affected by stress and indirectly by adherence [10]. For youths, the measures of metabolic control may be related to the intake quantity of insulin or hormonal changes caused as a result of puberty [25]. As the duration of the disease increases, youth find it upsetting and consider the regime to be difficult [12]. The role played in adherence behaviour by cognitive appraisal processes amongst adolescents was studied [26] and was found that poor adherence has high correlation with patients having negative perception about their bodies. These patients shared external attribution to negative events and had less perceived internal control over their health [26]. Poor metabolic control is also associated with worrying about the disease [27]. Patients attributing negative events to stable and internal causes were found to be suffering from learned helplessness and had low metabolic control [28].

Psychological interventions have helped in the lifestyle disorder management. Problems related to self-care, negative emotions, adaptation and quality of life have been tried to address by technology-intensive interventions along with psychological interventions like patient empowerment, coaching, cognitive behaviour therapy and behavioural skills [29]. A study on habit modification and drug adherence in obesity confirmed that adherence rates are better when cognitive behavioural therapy was combined with melatonin. An average of 80% adherence was achieved against 48% without the psychological techniques [30]. Another study on HIV patients revealed that multi-level interventions are required for addressing trauma and its negative impact on health and health-promoting behaviours [31]. A study in Japan used group cognitive-behavioural therapy (CBT) to reduce self-stigma and highlighted the relationship between cognitive restructuring and self-stigma. It was found that CBT helped to significantly reduce depression, anxiety and maladjusted cognitive bias [32]. Other studies also show a significant correlation between self-stigma and adherence to treatment [33].

A study on major depressive disorder revealed that medication adherence was strongly correlated with higher self-rated severity of illness, absence of personality pathology and lower perceived stigma [34]. In treatment of epilepsy patients, it was found that enhanced understanding of the disease by society, caregivers and patients was important [35]. In treatment of HIV patients, it is reported that even the adherent patients should be monitored [36]. Social support has been reported to positively affect adherence [37]. The presence of depressed mood and co-morbid illness, perception of health and medication effectiveness influenced more in adherence of COPD patients than their disease severity and demographic factors [38]. For children, it was found that parental supervision and disease severity help in adherence. Hence, partnership with patients and families about the treatment plan is useful [39].

Poor adherence is associated with several cognitive styles of patients: health control beliefs (internal and external), psychological reactance, medication scepticism and pharmacophobia [40]. A review conducted with 42 research papers suggests that additional research is required to address the gaps associated with cognitively impaired caregivers on patient's adherence, comparative effect of various techniques to increase adherence and role of pharmacists in finding and refining adherence [41]. In this study, cognitive barriers studied are denial, refusal and mistrust. These concepts are related but distinct.

Denial of an illness, severity of illness or need for medication is well known and has invited the attention of researchers. Adherence is directly affected by denial of disease as if there is no disease the question of taking medication does not arise at all. One of the earliest studies in this respect [42] began in 1938; it concluded that motivation to deny illness provided a unifying concept to cover various patterns of behaviour. Nathanson's [43] is the second oldest study on denial of disease. The study was undertaken to determine the incidence of denial of illness in patients in hemiplegia. Another study [44] traced the historical roots of denial of illness in order to understand how the construct of denial developed. The conclusion was that denial is a multidimensional construct; the use of denial can be both adaptive and maladaptive; and denial interacts with time to have a differential impact on outcomes. Measurement of denial is important. A few scholars have focused on it. One such study constructed a test instrument to assess the stage of acceptance and denial amongst patients with chronic kidney disease to identify the degree of denial amongst patients as a basis for intervention [45]. Ferrario [46] prepared a 24-question questionnaire and concluded that denial of negative emotions and resistance to change seem to contribute to a real expression of denial and conscious avoidance seems to constitute a further step in the process of cognitive-affective elaboration of the illness. Denial of disease has invited attention of a few researchers [47–49]. Denial is expected particularly from children [50]. Their study concludes that diabetes was very strongly associated with 'denial of disease' by children.

Refusal to take medication is another common phenomenon which leads to medical non-adherence. This too has invited interest of many a researcher. Baker et al. [51] studied the medication-related conflict born of service environment, physical environment and patient routines. They concluded that there is a possibility to ensure better medication adherence amongst patients in mental health wards by employing a more consensual approach to care. Most refusal studies are focussed on psychotic patients, children and geriatric patients. Studies of refusal to take medicine by diabetic patients too are found in the literature. Elisabeth et al. [52] identified refusal as a major reason for inability of physicians to prescribe medication to diabetic patients. Another study by Péres et al. [53] concludes that GPs do have to accept such refusals as a legitimate reason for not prescribing. Refusal to take insulin in Iran by patients with blood glucose levels is reported by Ghadiri-Anari et al. [54]. Their study confirms that insulin denial is prevalent. 77% of participant expressed unwillingness to take insulin if prescribed. Fear of injection is an important cause for insulin refusal amongst patients. Another study was undertaken for Korean diabetic patients [55]. The study also concluded that prevalence of insulin refusal in Korea is not low. A qualitative study focused upon providers' perspectives on cultural barriers and facilitators to diabetes self-management (DSM) in Arab Americans found that cultural reasons are at the root of refusal of medication by this ethnic minority.

Patient mistrust in healthcare system, professionals and medicines is age-old and in modern times has increased rather than decreasing [56]. Medical mistrust leads to lower utilization of healthcare and poorer management of health conditions including HIV, cancers and diabetes [57–64]. There are several diabetic-specific studies that explored mistrust of diabetic patients with medication. White et al. [65] and Hall et al. [66] studied Afro-American diabetic patients and concluded that one of the most prominent factors that crossed all barriers was medical mistrust. This conclusion also stems from the study of Shiyanbola et al. [67] and Ellis et al. [68].

This study explores these three variables in diabetic patients.

3 Methodology

As already suggested the objective of this study is to determine if significant difference in drug adherence behaviour with cognitive barriers (denial, refusal and mistrust) exists or not. It is also aimed to find if it is possible to categorize patients as high, moderate and low on adherence on the basis of their scores on denial, refusal and mistrust. The study is empirical, and data has been collected through a survey of diabetic patients at home. The questionnaire of the survey on cognitive barriers (Table 1) shows the description of variables. Three factors of cognitive behaviour associated with medication non-adherence associated with diabetic patients have been taken for the study: denial, refusal and mistrust. Five variables for each of the factor were framed, and survey was conducted amongst caregivers of diabetic patients. Screening questions included if denial of need of medication, refusal to take medicine or a mistrust for medication has affected the medication. Its face validity has been ensured by taking the views of four experts from the field on their appropriateness, clarity, objectivity and relevance. These have been measured on a 7-point Likert scale. 7 points were preferred over 5-point Likert scale as

Table 1 Description of variables

Factor	Description	Variable
Denial	Patient feels he/she has no disease	Denial1
	Patient says the disease is not severe	Denial2
	Patient says that disease has no severe effects	Denial3
	Patient says there is no need of medication	Denial4
	Patient says there is no need for regular medication	Denial5
Refusal	Patient refuses medication because of hopelessness	Refusal1
	Patient refuses medication because of his anger for being monitored	Refusal2
	Patient refuses medication because of his anger for being a victim of fate	Refusal3
	Patient refuses medication because he believes it is useless	Refusal4
	Patient refuses medication because of its perception as harmful	Refusal5
Mistrust	Patient does not take medicine because it is wastage of money	Mistrust1
	Patient does not take medicine because he believes alternate methods are better	Mistrust2
	Patient does not take medicine because it is not effective	Mistrust3
	Patient does not take medicine because it is not safe	Mistrust4
	Patient does not take medicine because of suffering of a known person	Mistrust5

respondents were more comfortable with the 7-point one. For adherence Brief Medication Questionnaire (BMQ) [69] has been used. It is a popular self-report tool for screening adherence. The scale comprises of five items that request patients on how the medication in the past week was taken by them. It was modified slightly to suit our needs. Permission to use the same was sought and granted. A diagrammatic representation of the methodology is given in Fig. 1.

The study was done with a sample size of 387 patient caregivers in India. The details of the demographic character of the respondents have been shown in Table 2. Convenience sampling method was used for the survey, and the mode adopted was online. Information was obtained for patients suffering from diabetes with the help of a structured questionnaire. The data obtained from the survey was analysed in SPSS version 26. The technique of ANOVA (CL 95%) and discriminant analysis was used for analysing the data. For discriminant analysis the scores were converted into three equal categories of low, moderate and high adherence.

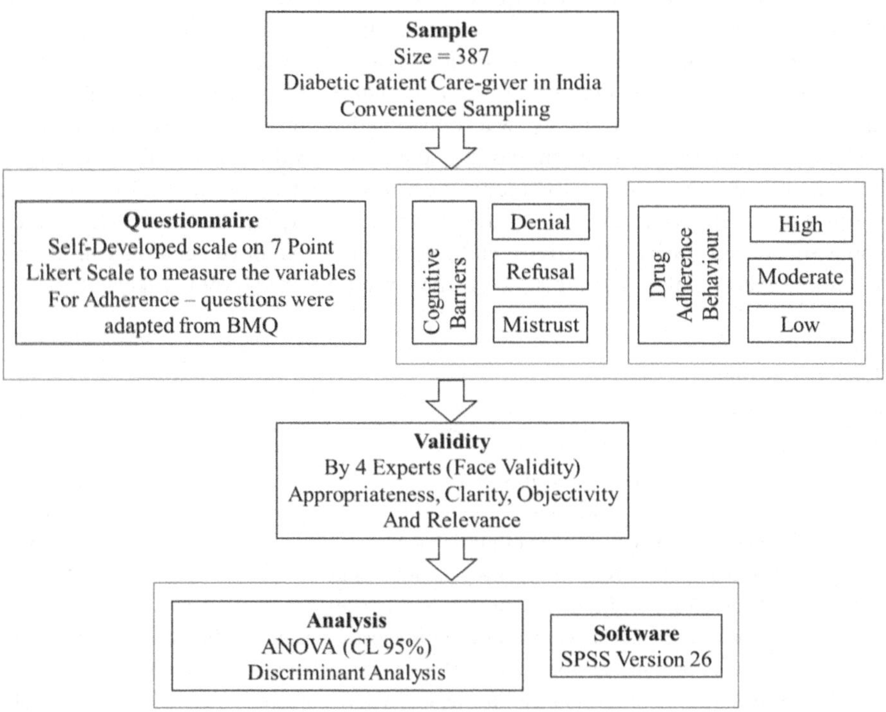

Fig. 1 Methodology

Table 2 Demographic character of respondents

Item	Category	Frequency	%
Gender	Male	217	56
	Female	170	44
Age	Under 45 year	*91*	23.5
	45–54 years	189	49
	55–64 years	88	23
	65 years or more	19	5.5
Education	Doctoral	*11*	2.8
	Masters	187	48.3
	Bachelor	189	48.8

4 Results and Discussion

Result and discussion have been been discussed in subsequent paragraphs and in the following sub-headings:

4.1 ANOVA (Denial vs. Adherence)
4.2 ANOVA (Refusal vs. Adherence)
4.3 ANOVA (Mistrust vs. Adherence)
4.4 Discriminant Analysis (Denial)
4.5 Discriminant Analysis (Refusal)
4.6 Discriminant Analysis (Mistrust)
4.7 Discriminant Analysis (Combined for All the Three Latent Constructs – Denial, Refusal and Mistrust

4.1 ANOVA (Denial vs. Adherence)

It was important to understand respondents with low, moderate and high adherence in their denial scores. To achieve this ANOVA has been used, and results have been shown in Table 3. From the table we find for the first variable on denial the mean values for low, moderate and high adherence are 6.03, 5.03 and 3.78. A significant difference is apparent. The same may be said for the other four denial variables under study.

It is important to explore if the differences are statistically significant. F values for the denial variables range from 47 to 61 and are high, thus indicating significant differences across low, moderate and high adherence exist. This is further confirmed by the p values which are near zero and much below the threshold level of 0.05. Values below 0.05 confirm that significant differences exist.

Table 3 ANOVA (denial vs. adherence)

	Adherence	Mean	Std. deviation	F value	p value
Denial1	Low	6.0308	1.04872	47	0
	Moderate	5.0398	1.61815		
	High	3.7818	1.7815		
Denial2	Low	5.6462	0.888	61	0
	Moderate	4.5622	1.68741		
	High	3.1091	1.60638		
Denial3	Low	6.1462	1.12829	45	0
	Moderate	4.9453	1.68285		
	High	4	1.66667		
Denial4	Low	5.0615	1.22477	55	0
	Moderate	4.2488	1.77421		
	High	2.3636	1.65958		
Denial5	Low	5.8077	0.94086	57	0
	Moderate	4.7413	1.62257		
	High	3.4182	1.62948		

Table 4 ANOVA (refusal vs. adherence)

	Adherence	Mean	Std. deviation	F value	p value
Refusal1	Low	6.1385	0.83298	67	0
	Moderate	4.8706	1.61962		
	High	3.6182	1.74792		
Refusal2	Low	5.6462	0.93892	75	0
	Moderate	4.5522	1.60577		
	High	2.9273	1.51357		
Refusal3	Low	5.8538	1.25205	59	0
	Moderate	4.3781	1.79062		
	High	3.2727	1.67171		
Refusal4	Low	5.6769	1.23399	37	0
	Moderate	4.9005	1.75785		
	High	3.4	2.03306		
Refusal5	Low	5.7	1.09721	60	0
	Moderate	4.4627	1.72622		
	High	3.0909	1.68075		

4.2 ANOVA (Refusal vs. Adherence)

As the second step, the refusal values against low, moderate and high adherence have been investigated, and the results shared in Table 4. It is to be observed that for the first variable measuring refusal (Refusal1) has mean values 6.13, 4.87 and 3.62 for low, moderate and high adherence. It holds true for other refusal variables, namely, Refusal2, Refusal3, Refusal4 and Refusal5. Differences are rather perceptible across adherence, and this needs statistical confirmation. Corresponding *F* values are found to range between 37 and 67. This clearly establishes that significant differences in refusal values across low, moderate and high adherence exist. This confirmation is echoed by the corresponding *p* value which is zero, well below 0.05.

4.3 ANOVA (Mistrust vs. Adherence)

Third step in this direction explores the differences in the scores of all the variables of mistrust across low to high adherence. Table 5 provides an understanding of the same. Mean values of all the five variables of mistrust are found to differ visibly. Are these differences statistically significant? To answer this question, corresponding *F* values provide interesting insight. Again, we find that *F* values are rather high spread between 47 and 64. As the *p* value for all the factors are <0.05, it is statistically established that the means are different. Significant influence of mistrust on low to high adherence is therefore settled. This is corroborated by corresponding *p* values which are well below 0.05

Table 5 ANOVA (mistrust vs. adherence)

	Adherence	Mean	Std. deviation	*F* value	*p* value
Mistrust1	Low	6.0769	0.99312	64	0
	Moderate	4.8507	1.57722		
	High	3.5636	1.78225		
Mistrust2	Low	5.5769	1.04076	55	0
	Moderate	4.5721	1.62666		
	High	3.1273	1.73263		
Mstrust3	Low	5.6846	1.39231	49	0
	Moderate	4.3085	1.82876		
	High	3.2	1.68215		
Mistrust4	Low	5.3692	1.2272	64	0
	Moderate	4.3433	1.61758		
	High	2.6727	1.57591		
Mistrust5	Low	5.7692	1.05315	47	0
	Moderate	4.8209	1.56133		
	High	3.5455	1.79318		

4.4 Discriminant Analysis (Denial)

After ANOVA establishes the significant differences in denial, refusal and mistrust values across low to high adherence, the next logical step is to check if this could be useful for classifying respondents with low, moderate and high adherence. Discriminant analysis is a useful tool in this respect, and it was taken recourse to. Before this analysis is undertaken, the suitability with given data needs to be understood. The indicators in this respect are eigenvalue and Wilks' lambda which should be low and high, respectively, to indicate appropriate discrimination. In this case, the eigenvalue is 0.452 which is low. Such a low value is associated with weak function. The value of Wilks' lambda is 0.684, which is relatively high and implies weak discrimination. Despite the weak functions, discriminant analysis still appeared possible. As a next step, structure matrix was prepared to understand the relative influence of individual denial variables. From the structure matrix, it is to be seen that highest influence is on discrimination, and subsequent classification is exerted by Denial5 and Denial2. However, as all values are above 0.7, all exert discriminant character in this respect.

The next step involves classifying the low, moderate and high adherent respondents and checking the correctness of this classification. Table 6 depicts the classification results of the variable denial. In this study, only 62% of original grouped cases were correctly classified for the variable denial. Out of 133 low adherence patients, 107 have been identified correctly, and 26 have been grouped wrongly. Numbers of correct and wrong classification for moderate and high respondents are 118 out of 235 and 16 out of 18, respectively. Is 62% classification acceptable? With three categories correct classification above 50% is considered acceptable. However, only 62% correct grouping is not very encouraging (Table 7).

4.5 Discriminant Analysis (Refusal)

For discriminant analysis of latent construct refusal, the corresponding eigenvalue is 0.606 which is still low. Such a low value is associated with weak function as already stated. The value of Wilks' lambda is 0.612, which again is high and implies weak discrimination. The structure matrix for refusal has been depicted in Table 8.

Table 6 Structure matrix for denial

Function	
Denial5	0.921*
Denial2	0.865*
Denial1	0.838*
Denial3	0.768*
Denial4	0.732*

*Discriminating capacity is highest with variable Denial5, followed by variables Denial 2, Denial 1, Denial 3 and Denial 4

Table 7 Classification results (denial)

			Predicted group membership			Total
Adherence			1	2	3	
Original	Count	Low	107	23	3	133
		Moderate	74	118	43	235
		High	0	2	16	18
	%	Low	80.5	17.3	2.3	100
		Moderate	31.5	50.2	18.3	100
		High	0	11.1	88.9	100
62% of original grouped cases correctly classified						

Table 8 Structure matrix – refusal

Function	
Refusal2	0.859*
Refusal3	0.851*
Refusal1	0.814*
Refusal5	0.782*
Refusal4	0.565

*Discrimination is highest with variable Refusal2, followed by variables Refusal3, Refusal1, Refusal5 and Refusal 4

Variables Refusal2 and Refusal3 emerge as highly discriminating compared to Refusal1, Refusal5 and Refusal4.

Table 9 depicts the classification results of the variable refusal. Out of the total 133 patients, 97 (72.9%) of the patients with low adherence could be predicted correctly for the variable refusal. Patients belonging to the category of moderate adherence could also be predicted, and the percentage for it was 59.6%. Out of 235 patients belonging to this category, barely 140 could be predicted accurately. For patients with high adherence, the classification was 77.8% accurate. In this study, 65% of original grouped cases were correctly classified for the variable refusal. The result is acceptable though not very promising.

4.6 *Discriminant Analysis (Mistrust)*

Latent construct mistrust has been analysed through discriminant analysis as the next step. The eigenvalue is 0.621, which is very low. Such a low value is associated with weak function. The value of Wilks' lambda is 0.602, which is high and implies weak discrimination. Table 10, reflecting the structure matrix, shows that most significant discriminating variables in this respect are Mistrust1 and Mistrust2.

Table 9 Classification results (refusal)

Adherence			Predicted group membership 1	2	3	Total
Original	Count	Low	97	36	0	133
		Moderate	51	140	44	235
		High	0	4	14	18
	%	Low	72.9	27.1	0	100
		Moderate	21.7	59.6	18.7	100
		High	0	22.2	77.8	100

a. 65% of original grouped cases correctly classified.

Table 10 Structure matrix (mistrust)

Function	
Mistrust1	0.86
Mistrust2	0.758
Mistrust3	0.741
Mistrust4	0.73
Mistrust5	0.727

Table 11 depicts the classification results of the variable mistrust. Out of the total 133 patients, 80.5% of the patients with low adherence could be predicted correctly. The number for which was 107. Patients belonging to the category of moderate adherence could also be predicted, and the percentage for it was 55%. Out of 235 patients belonging to this category, 129 could be predicted accurately. For patients with high adherence, the classification was 88.9% accurate. In this study, 65.3% of original grouped cases were correctly classified for the variable mistrust. 65.3% is acceptable number but again lacks the decisiveness.

4.7 *Discriminant Analysis (Combined for All the Three Latent Constructs – Denial, Refusal and Mistrust)*

Having failed to have good discrimination with the three latent constructs under this study to a significant extent, it was decided to perform a discriminant analysis for all the three latent constructs of this study. This has been done on the premise that the three latent constructs do not correctly or wrongly identify the same respondents with low to high adherence. The eigenvalue obtained is 3.365, which is high. Such a high value is associated with strong function. The value of Wilks' lambda is 0.22, which is low and implies strong discrimination. This provides a strong basis for performing discriminant analysis.

Table 12 depicts the classification results for low, moderate and high adherence on the basis of denial, refusal and mistrust taken together. Out of the total 133 patients, 132 low adherent respondents making 99.2% could be predicted correctly.

Table 11 Classification results (mistrust)

			Predicted group membership			Total
Adherence			1	2	3	
Original	Count	Low	107	25	1	133
		Moderate	64	129	42	235
		High	0	2	16	18
	%	Low	80.5	18.8	0.8	100
		Moderate	27.2	55	17.9	100
		High	0	11.1	88.9	100

a. 65.3% of original grouped cases correctly classified

Table 12 Classification results (denial + refusal + mistrust)

			Predicted group membership			Total
Adherence			1	2	3	
Original	Count	1	132	1	0	133
		2	10	223	2	235
		3	0	0	18	18
	%	1	99.2	0.8	0	100
		2	4.3	94.9	0.9	100
		3	0	0	100	100

96.6% of original grouped cases correctly classified

Patients belonging to the category of moderate adherence could also be predicted, and the percentage for it was 94.9%. Out of 235 patients belonging to this category, 223 could be predicted accurately. For patients with high adherence, the classification was 100% accurate. In this study, 96.6% of original grouped cases were correctly classified.

The results confirm our assumption that denial, refusal and mistrust are distinct latent constructs, and these classify adherence behaviour differently indicating a discriminant validity of all the three.

5 Conclusion

It is common for patients to poorly adhere to the medication regime prescribed by their physician. It leads to worsening of disease, increased healthcare costs and even death. Professionals associated with healthcare and treatment of patients should look for enhancing the medication adherence rate of patients by emphasizing the value of a patient's regimen, simplifying it with customization as per the patient's lifestyle [70]. They should be aware of the factors of non-adherence for the patient [71]. This understanding is essential to develop interventions that will improve adherence to therapeutic regimens [72]. The present study tries to find the adherence level of diabetic patients into three different categories (low, moderate and high adherence). Three factors of cognitive behaviour associated with medication non-adherence were identified for the study: denial, refusal and mistrust. The ANOVA shows that the three factors taken for the study are different (means are not equal as $p < 0.05$ at 95% confidence level). The discriminant analysis performed for each factor shows that there is weak but significant discriminant within the factors (63–66%), but on performing discriminant analysis with all the factors, it was found that there is strong and significant discrimination 96.6%.

Studies report that youths who lost control of their blood sugar level and who perceived little internal control had higher probability of non-adherence. They failed to develop a cognitive alignment to mastery and used the metre reading information to figure out their adherence problems [26]. Literature suggests that a patient's thought process about a situation affects his/her emotions and behaviour. A patient having distorted and negative thinking is more likely to have negative feelings, and the behaviour will be affected accordingly [73].

Patient's attitude towards drugs [74], cost [75], patient-clinician relationship [76] and knowledge about the disease [77] has a major role in drug adherence. Mapping patients' everyday activities and associating them with medications help to increase adherence [78]. A study on hypertension and cardiovascular disease in blacks suggests eight practical approaches to addressing medication adherence: (1) cognitive-based behaviour, (2) consumer-directed healthcare, (3) patient engagement strategies, (4) cardiac rehabilitation, (5) patient portals, (6) smart apps and text messages, (7) pharmacist-led engagement and (6) digital pillboxes [79]. A study in Africa reveals that patients suffering from type 2 diabetes cited the reasons for intentional non-adherence as perception of medicines, concern about side effects, fear and frustration of medication, perception of the disease, information and access to medicines. The study highlighted the reasons for adherence were patients' importance for being alive, performing their family and social duties and their beliefs in physician's instruction [67].

Technology-intensive intervention methods have proved to increase the adherence rates subsequently. As there is no "gold standard" for measuring adherence, a combination of methods [80] may be used for assessing the adherence rate of patients, measure the distinct behaviours involved in the schedule [10] and frame strategies to improve the adherence rate [81].

The results provide an empirical basis for predicting the adherence behaviour of patients on the basis. These have implications for healthcare providers, healthcare regulators and patients and their family members. Low adherence is a poor behaviour and affects healthcare outcomes. It can be identified correctly for patients reporting denial, refusal and mistrust. This study should help in developing a mechanism to identify low adherent patients and then evolve ways and means to counter cognitive barriers.

References

1. Rolnick, S.J., Pawloski, P.A., Hedblom, B.D., Asche, S.E., Bruzek, R.J.: Patient characteristics associated with medication adherence. Clin. Med. Res. **11**, 54–65 (2013)
2. Nichol, M.B., Venturini, F., Sung, J.C.Y.: A critical evaluation of the methodology of the literature on medication compliance. Ann. Pharmacother. **33**, 531–540 (1999)
3. Schlenk, E.A., Burke, L.E., Rand, C.: Behavioral strategies to improve medication-taking compliance. In: Compliance in Healthcare and Research, pp. 57–70. Futura Publishing Company, Inc., Armonk (2001)
4. Schroeder, K., Fahey, T., Ebrahim, S.: How can we improve adherence to blood pressure-lowering medication in ambulatory care? Systematic review of randomized controlled trials. Arch. Intern. Med. **164**, 722–732 (2004)
5. Roebuck, M.C., Liberman, J.N., Gemmill-Toyama, M., Brennan, T.A.: Medication adherence leads to lower health care use and costs despite increased drug spending. Health Aff. **30**, 91–99 (2011)
6. WHO I Adherence to long-term therapies: evidence for action. WHO (World Health Organization, 2015)
7. Sabaté, E., Organization, W. H., & others: Adherence to long-term therapies: policy for action: meeting report, 4–5 June 2001, (2001)
8. McDonnell, P.J., Jacobs, M.R., Monsanto, H.A., Kaiser, J.M.: Hospital admissions resulting from preventable adverse drug reactions. Ann. Pharmacother. **36**, 1331–1336 (2002)
9. Christie, M.J., French, D., Weatherstone, L., West, A.: The patients' perceptions of chronic disease and its management: psychosomatics, holism and quality of life in contemporary management of childhood asthma. Psychother. Psychosom. **56**, 197–203 (1991)
10. Johnson, S.B.: Insulin-dependent diabetes mellitus in childhood (1995)
11. Kendall, P.C.: Anxiety disorders in youth (2012)
12. Kovacs, M., et al.: Psychological functioning of children with insulin-dependent diabetes mellitus: a longitudinal study. J. Pediatr. Psychol. **15**, 619–632 (1990)
13. Cramer, J.A., et al.: Medication compliance and persistence: terminology and definitions. Value Health. **11**, 44–47 (2008)
14. Osterberg, L., Blaschke, T.: Adherence to medication. N. Engl. J. Med. **353**, 487–497 (2005)
15. Gossec, L., Tubach, F., Dougados, M., Ravaud, P.: Reporting of adherence to medication in recent randomized controlled trials of 6 chronic diseases: a systematic literature review. Am. J. Med. Sci. **334**, 248–254 (2007)
16. Ho, P.M., Bryson, C.L., Rumsfeld, J.S.: Medication adherence: its importance in cardiovascular outcomes. Circulation. **119**, 3028–3035 (2009)
17. MacLaughlin, E.J., et al.: Assessing medication adherence in the elderly: which tools to use in clinical practice? Drugs Aging. **22**, 231–255 (2005)
18. Sabaté, E., Sabaté, E., & others: Adherence to long-term therapies: evidence for action. (World Health Organization, 2003)
19. Charles, H., Good, C.B., Hanusa, B.H., Chang, C.C.H., Whittle, J.: Racial differences in adherence to cardiac medications. J. Natl. Med. Assoc. **95**, 17–27 (2003)
20. Hyre, A.D., Krousel-Wood, M.A., Muntner, P., Kawasaki, L., DeSalvo, K.B.: Prevalence and predictors of poor antihypertensive medication adherence in an urban health clinic setting. J. Clin. Hypertens. **9**, 179–186 (2007)
21. Mochari, H., Ferris, A., Adigopula, S., Henry, G., Mosca, L.: Cardiovascular disease knowledge, medication adherence, and barriers to preventive action in a minority population. Prev. Cardiol. **10**, 190–195 (2007)
22. Bardel, A., Wallander, M.A., Svärdsudd, K.: Factors associated with adherence to drug therapy: a population-based study. Eur. J. Clin. Pharmacol. **63**, 307–314 (2007)
23. Rieckmann, N., et al.: Course of depressive symptoms and medication adherence after acute coronary syndromes. An electronic medication monitoring study. J. Am. Coll. Cardiol. **48**, 2218–2222 (2006)

24. Farrell, S.P., Hains, A.A., Davies, W.H., Smith, P., Parton, E.: The impact of cognitive distortions, stress, and adherence on metabolic control in youths with type 1 diabetes. J. Adolesc. Health. **34**, 461–467 (2004)
25. Reid, P., Appleton, P.: Insulin dependent diabetes mellitus: regimen adherence in children and young people. Irish J. Psychol. **12**, 17–32 (1991)
26. Bennett Murphy, L.M., Thompson, R.J., Morris, M.A.: Adherence behavior among adolescents with type I insulin-dependent diabetes mellitus: the role of cognitive appraisal processes. J. Pediatr. Psychol. **22**, 811–825 (1997)
27. Liakopoulou, M., Korvessi, M., Dacou-Voutetakis, C.: Personality characteristics, environmental factors and glycemic control in adolescents with diabetes. Eur. Child Adolesc. Psychiatry. **1**, 82–88 (1992)
28. Kuttner, M.J., Delamater, A.M., Santiago, J.V.: Learned helplessness in diabetic youths. J. Pediatr. Psychol. **15**, 581–594 (1990)
29. Sudhir, P.M.: Advances in psychological interventions for lifestyle disorders: overview of interventions in cardiovascular disorder and type 2 diabetes mellitus. Curr. Opin. Psychiatry. **30**, 346–351 (2017)
30. Sosa, V.R.A., González, M.D.L.L.S., Bandala, C., Vargas de León, C., Lara Padilla, E., Marín Soto, M.D., ... & Murillo Tovar, M.M: Efficacy of Cognitive Behavioral Therapy for Habit Modification and Drug Adherence in Obesity. Acta de investigación psicológica, **10**(1), 49–56 (2020).
31. Sales, J.M., Swartzendruber, A., Phillips, A.L.: Trauma-informed HIV prevention and treatment. Curr. HIV/AIDS Rep. **13**, 374–382 (2016)
32. Shimotsu, S., et al.: Effectiveness of group cognitive-behavioral therapy in reducing self-stigma in Japanese psychiatric patients. Asian J. Psychiatr. **10**, 39–44 (2014)
33. Kamaradova, D., et al.: Connection between self-stigma, adherence to treatment, and discontinuation of medication. Patient Prefer. Adherence. **10**, 1289–1298 (2016)
34. Sirey, J.A., et al.: Perceived stigma and patient-rated severity of illness as predictors of antidepressant drug adherence. Psychiatr. Serv. **52**, 1615–1620 (2001)
35. Garnett, W.R.: Antiepileptic drug treatment: Outcomes and adherence. Pharmacotherapy. **20**, 191S–199S. Pharmacotherapy Publications Inc. (2000)
36. Iacob, S.A., Iacob, D.G., Jugulete, G.: Improving the adherence to antiretroviral therapy, a difficult but essential task for a successful HIV treatment—clinical points of view and practical considerations. Front. Pharmacol. **8**, 831 (2017)
37. Vyavaharkar, M., et al.: Social support, coping, and medication adherence among HIV-positive women with depression living in rural areas of the southeastern United States. AIDS Patient Care STDS. **21**, 667–680 (2007)
38. Khdour, M.R., Hawwa, A.F., Kidney, J.C., Smyth, B.M., McElnay, J.C.: Potential risk factors for medication non-adherence in patients with chronic obstructive pulmonary disease (COPD). Eur. J. Clin. Pharmacol. **68**, 1365–1373 (2012)
39. Zindani, G.N., Streetman, D.D., Streetman, D.S., Nasr, S.Z.: Adherence to treatment in children and adolescent patients with cystic fibrosis. J. Adolesc. Health. **38**, 13–17 (2006)
40. De las Cuevas, C., de Leon, J.: Reviving research on medication attitudes for improving pharmacotherapy: focusing on adherence. Psychother. Psychosom. **86**, 73–79 (2017)
41. Hudani, Z.K., Rojas-Fernandez, C.H.: A scoping review on medication adherence in older patients with cognitive impairment or dementia. Res. Social Adm. Pharm. **12**, 815–829 (2016)
42. Weinstein, E.A., Kahn, R.L.: Denial of Illness: Symbolic and Physiological Aspects. Thomas, Springfield (1955)
43. Nathanson, M., Bergman, P.S., Gordon, G.G.: Denial of illness: its occurrence in one hundred consecutive cases of hemiplegia. Arch. Neurol. Psychiatry. **68**, 380–387 (1952)
44. Kortte, K.B., Wegener, S.T.: Denial of illness in medical rehabilitation populations: theory, research, and definition. Rehabil. Psychol. **49**, 187–199 (2004)
45. Gagani, A., Gemao, J., Relojo, D., Pilao, S.J.: The dilemma of denial: acceptance and individual coping among patients with chronic kidney diseases. J. Educ. Sci. Psychol. **6**(2), (2016)

46. Ferrario, S.R., et al.: Illness denial questionnaire for patients and caregivers. Neuropsychiatr. Dis. Treat. **13**, 909–916 (2017)
47. Garay-Sevilla, M.E., Malacara, J.M., Gutierrez-Roa, A., Gonzalez, E.: Denial of disease in type 2 diabetes mellitus: its influence on metabolic control and associated factors. Diabet. Med. **16**, 238–244 (1999)
48. Broom, D., Whittaker, A.: Controlling diabetes, controlling diabetics: moral language in the management of diabetes type 2. Soc. Sci. Med. **58**, 2371–2382 (2004)
49. Dhesi, S., Ellenbogen, J.: Multidisciplinary teamwork: collaborating on diabetes. Br. Dent. J. **222**, 406 (2017)
50. Saoji, N., et al.: The relationship of type 1 diabetes self-management barriers to child and parent quality of life: a US cross-sectional study. Diabet. Med. **35**, 1523–1530 (2018)
51. Baker, J.A., Bowers, L., Owiti, J.A.: Wards features associated with high rates of medication refusal by patients: a large multi-centred survey. Gen. Hosp. Psychiatry. **31**, 80–89 (2009)
52. Ab, E., Denig, P., Van Vliet, T., Dekker, J.H.: Reasons of general practitioners for not prescribing lipid-lowering medication to patients with diabetes: a qualitative study. BMC Fam. Pract. **10**, 1–7 (2009)
53. Péres, D.S., Dos Santos, M.A., Zanetti, M.L., Ferronato, A.A.: Difficulties of diabetic patients in the illness control: feelings and behaviors. Rev. Lat. Am. Enfermagem. **15**, 1105–1112 (2007)
54. Ghadiri-Anari, A., Fazaelipoor, Z., Mohammadi, S.M.: Insulin refusal in Iranian patients with poorly controlled type 2 diabetes mellitus. Acta Med. Iran. **51**, 567–571 (2013)
55. Song, Y., et al.: The prevalence of insulin refusal and psychological insulin resistance among Korean patients with type 2 diabetes mellitus. Ann. Transl. Med. **7**, 760–760 (2019)
56. Jaiswal, J., Halkitis, P.N.: Towards a more inclusive and dynamic understanding of medical mistrust informed by science. Behav. Med. **45**, 79–85 (2019)
57. Pellowski, J.A., Price, D.M., Allen, A.M., Eaton, L.A., Kalichman, S.C.: The differences between medical trust and mistrust and their respective influences on medication beliefs and ART adherence among African-Americans living with HIV. Psychol. Health. **32**, 1127–1139 (2017)
58. Cahill, S., et al.: Stigma, medical mistrust, and perceived racism may affect PrEP awareness and uptake in black compared to white gay and bisexual men in Jackson, Mississippi and Boston, Massachusetts. AIDS Care Psychol. Socio Med. Asp. AIDS/HIV. **29**, 1351–1358 (2017)
59. Babushkina, E.A., Belokopytova, L.V., Grachev, A.M., Meko, D.M., Vaganov, E.A.: Variation of the hydrological regime of Bele-Shira closed basin in Southern Siberia and its reflection in the radial growth of Larix sibirica. Reg. Environ. Chang. **17**, 1725–1737 (2017)
60. LaVeist, T.A., Nickerson, K.J., Bowie, J.V.: Attitudes about racism, medical mistrust, and satisfaction with care among African American and white cardiac patients. Med. Care Res. Rev. **57**, 146–161 (2000)
61. Benkert, R., Peters, R.M., Clark, R., Keves-Foster, K.: Effects of perceived racism, cultural mistrust and trust in providers on satisfaction with care. J. Natl. Med. Assoc. **98**, 1532–1540 (2006)
62. Kinlock, B.L., et al.: High levels of medical mistrust are associated with low quality of life among black and white men with prostate cancer. Cancer Control. **24**, 72–77 (2017)
63. Lee, H., Park, S.A.: Third-person effect and pandemic flu: the role of severity, self-efficacy method mentions, and message source. J. Health Commun. **21**, 1244–1250 (2016)
64. Kolar, S.K., et al.: Human papillomavirus vaccine knowledge and attitudes, preventative health behaviors, and medical mistrust among a racially and ethnically diverse sample of college women. J. Racial Ethn. Health Disparities. **2**, 77–85 (2015)
65. White, R.O., et al.: Perceptions of provider communication among vulnerable patients with diabetes: influences of medical mistrust and health literacy. J. Health Commun. **21**, 127–134 (2016)
66. Hall, G.L., Heath, M.: Poor medication adherence in African Americans is a matter of trust. J. Racial Ethn. Health Disparities 1–16 (2020). https://doi.org/10.1007/s40615-020-00850-3

67. Shiyanbola, O.O., Brown, C.M., Ward, E.C.: "I did not want to take that medicine": African-Americans' reasons for diabetes medication nonadherence and perceived solutions for enhancing adherence. Patient Prefer. Adherence. **12**, 409–421 (2018)
68. Ellis, J.L., et al.: Factors related to medication self-management in African American older women. Res. Gerontol. Nurs. **12**, 71–79 (2019)
69. Svarstad, B.L., Chewning, B.A., Sleath, B.L., Claesson, C.: The brief medication questionnaire: a tool for screening patient adherence and barriers to adherence. Patient Educ. Couns. **37**, 113–124 (1999)
70. Zygmunt, A., Olfson, M., Boyer, C.A., Mechanic, D.: Interventions to improve medication adherence in schizophrenia. Am. J. Psychiatry. **159**, 1653–1664 (2002)
71. Daley, D.J., Myint, P.K., Gray, R.J., Deane, K.H.O.L.: Systematic review on factors associated with medication non-adherence in Parkinson's disease. Parkinsonism Relat. Disord. **18**, 1053–1061 (2012)
72. García, P.R., Côté, J.K.: Factors affecting adherence to antiretroviral therapy in people living with HIV/AIDS. J. Assoc. Nurses AIDS Care. **14**, 37–45 (2003)
73. Thomas, A.M., Peterson, L., Goldstein, D.: Problem solving and diabetes regimen adherence by children and adolescents with IDDM in social pressure situations: a reflection of normal development. J. Pediatr. Psychol. **22**, 541–561 (1997)
74. Ghaffari-Nejad, A., Mashayekhi, A., Mazhari, S., Ghayoumi, A.: Factors associated with drug non-adherence after discharge among patients with mood disorders hospitalized in Shahid Beheshti Hospital in Kerman: a prospective study. Subst. Abus. **57**, 38 (2015)
75. Wang, Y., et al.: Drug non-adherence and reasons among multidrug-resistant tuberculosis patients in Guizhou, China: a cross-sectional study. Patient Prefer. Adherence. **13**, 1641–1653 (2019)
76. Pucci, M., Martin, U.: Detecting non-adherence by urine analysis in patients with uncontrolled hypertension: rates, reasons and reactions. J. Hum. Hypertens. **31**, 253–257 (2017)
77. Jankowska-Polańska, B., Uchmanowicz, I., Dudek, K., Mazur, G.: Relationship between patients' knowledge and medication adherence among patients with hypertension. Patient Prefer. Adherence. **10**, 2437–2447 (2016)
78. Mechta Nielsen, T., Frøjk Juhl, M., Feldt-Rasmussen, B., Thomsen, T.: Adherence to medication in patients with chronic kidney disease: a systematic review of qualitative research. Clin. Kidney J. **11**, 513–527 (2018)
79. Ferdinand, K.C., et al.: Disparities in hypertension and cardiovascular disease in blacks: the critical role of medication adherence. J. Clin. Hypertens. **19**, 1015–1024 (2017)
80. Fisher, B.G., Acerini, C.L.: Understanding the growth hormone therapy adherence paradigm: a systematic review. Horm. Res. Paediatr. **79**, 189–196 (2013)
81. Tucker, C.M., et al.: Associations with medication adherence among ethnically different pediatric patients with renal transplants. Pediatr. Nephrol. **17**, 251–256 (2002)

Study on the Effects of Startup Ecosystem Components on Overcoming Death-Valley in SMEs – Focusing on the Mediating Effect of the Win-Win Platform

Tae-Ho You, Yen-Yoo You, and In-Chae Park

1 Introduction

South Korea is a marginal country in the startup world. The number of unicorn companies is 11, ranked joint fifth in the world behind the United States, China, The United Kingdom and India, followed by Germany [1]. According to a 2017 survey of 55 cities in the world by Startup Genome, "Value of the Start-up Ecosystem by City" showed 264 points in Silicon Valley, 131 in Beijing, 22 in China, and 11 in Singapore, while Seoul scored 2.4 points [2]. The first formula for Silicon Valley's "innovation equation" is "minimizing regulation." However, due to government regulations, Korea does not allow startups to grow, and eventually, deacon companies do not exist. When asked first "in the past year are there regulatory slowdowns in business?" 47.5% of companies answered yes. Second, Korea's finance is not structured because it is mainly based on real estate, which is a typical tangible asset. In order for a startup to grow into a unicorn, large capital must be introduced through finance to create a virtuous cycle structure. The third is the issue of human resources. In order to grow into a unicorn company, creative talents are needed, and Korean education methods are difficult to come up with creativity due to entrance exams and memorization-oriented education. Therefore, there is an extreme shortage of people who are progressively challenging. These problems have only a survival rate of 39% for Korean startups for over 3 years. In addition, after 13 years of IPO and M&A, the probability of success is only 1.3%. Therefore, Korea should make policy efforts to build a healthy startup ecosystem so that startups can become the subject selves of innovation growth. This research aims to realize the sustainable management of startups by re-examining the components of the entrepreneurial ecosystem

T.-H. You (✉) · Y.-Y. You · I.-C. Park
Department of Smart Convergence Consulting, Hansung University, Seoul, South Korea
e-mail: threey0818@hansung.ac.kr; ipark@hansung.ac.kr

S. R. Samanta et al. (eds.), *Cognitive Computing for Risk Management*,
EAI/Springer Innovations in Communication and Computing,
https://doi.org/10.1007/978-3-030-74517-2_4

that enhance the overcoming of death-valley through prior research through a win-win platform [3].

2 Materials and Methods

2.1 *Concept of Startup Ecosystem*

The ecosystem focuses on the fact that various elements are organized systematically and equilibrium, rather than simply uniting them into units that refer to an integrated system between the environment and the biological population. In social sciences, it is used as a term for social ecosystems, corporate ecosystems, etc., by combining evolutionary meanings, and the concept of the business ecosystem of startup companies is interpreted as an expansion of the concept of the ecosystem [4]. The study presented three components of the entrepreneurial ecosystem: government policy, human resources, and financial support.

2.1.1 Concept of Policy

Korea's startup support policy supports companies in the 3–7 years of startup as a "startup leap package" under the Ministry of Small and Medium-Sized Venture Enterprises, and the main contents include commercialization funding, export support, product improvement support, investment support, prototyping support, consultation, service and information, and marketing cost support [5].

2.1.2 Concept of Human Resource

In order for Korean startups to grow into unicorn companies, it is essential to have creative people. Creativity should be nurtured through education, and education in Korea is thoroughly entrance and memorization. There's a situation in which progressive talent is extremely scarce [6].

2.1.3 Concept of Finance

In order for a startup to grow into a unicorn company, large capital must be introduced through financing, and Korea is a structure that is difficult to attract capital from. When we review the company's development stage on a financing basis, the startup stage requires initial demand funds, funds for commercialization support at the growth stage, and funds for sustainable growth at the maturity stage [7].

2.2 Concept of Win-Win Platforms

"Win-win platform" is an innovation platform that is configured to exchange ideas, knowledge, information, and opinions necessary to pursue coexistence, transform ingesting existing vertical cooperation networks into a horizontal structure system for collaboration and symbiosis among participants, and establish an operating entity that coordinates and manages autonomous cooperation between participants and promotes the growth of the platform as a whole [8].

2.3 Concept of Overcoming Death-Valley

It is common for startup companies to experience a lack of funds in the process of moving to the proof of concept stage after the development of original technology, in the case of medium-term enterprises, but successful in the development of prototypes, facing the death-valley which has difficulty in securing funds in the market-ready product, and in order to overcome this, a virtuous cycle of the startup ecosystem must be created with a variety of support [9].

2.4 Win-Win Platform and Death-Valley Overcoming Relationship

In order to increase productivity and sustainability activities within the business ecosystem, stability and productivity can be expanded when a mutual platform belonging to small- and medium-sized enterprises is established in a virtuous cycle structure, and the company's death-valley can be overcome [10].

2.5 The Research Model and Hypothesis

2.5.1 Research Model

The purpose of this study is to find the optimal combination of the entrepreneurial ecosystem components and the win-win platform to analyze the impact of the entrepreneur's overcoming of death-valley. To this end, government policies, human resources, and financial assistance were used as independent variables. The research model was completed by setting the win-win platform as a dependent variable between the entrepreneurial ecosystem components and the parameters of overcoming death-valley. It is presented as the following (Fig. 1) model of this study.

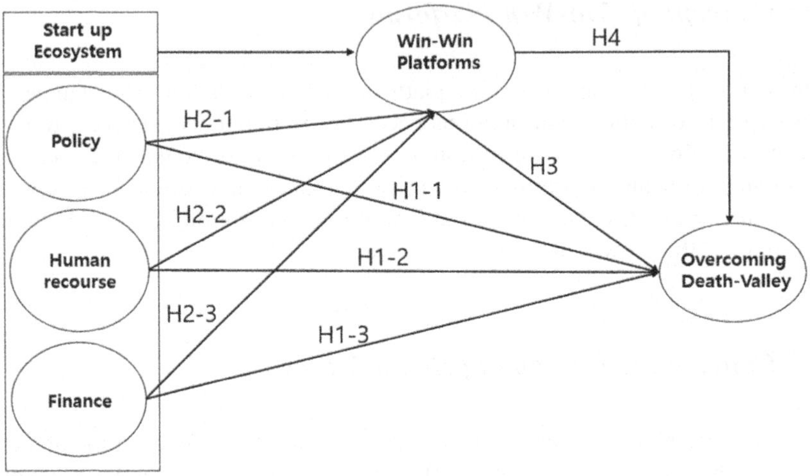

Fig. 1 Research model

2.5.2 Research Hypothesis

Figure 1 is based on the following research hypothesis regarding the impact of the relationship between the entrepreneurial ecosystem components and the win-win platform to overcome death-valley.

H1-1 Policy startup ecosystem will have a positive effect on overcoming death-valley.
H1-2 Human resource startup ecosystem will have a positive effect on overcoming death-valley.
H1-3 Finance startup ecosystem will have a positive effect on overcoming death-valley.
H2-1 Policy startup ecosystem will have a positive effect on the win-win platforms.
H2-2 Human resource startup ecosystem will have a positive effect on the win-win platforms.
H2-3 Finance startup ecosystem will have a positive effect on the win-win platforms.
H3 Win-win platforms will have a positive effect on overcoming death-valley.
H4-1 The mediating effect of win-win platforms has a positive effect on the relationship between policy and overcoming death-valley.
H4-2 The mediating effect of win-win platforms has a positive effect on the relationship between human resource and overcoming death-valley.
H4-3 The mediating effect of win-win platforms has a positive effect on the relationship between finance and overcoming death-valley.

2.5.3 Operational Definition of Variables

The operational definitions of each variable are summarized as follows (Table 1).

Table 1 Operational definition of variables

Evaluation	Measurement variable	Operational definition	Configure questionnaire	Researcher
Startup ecosystem	Policy	Korea's entrepreneurship support policy is the "Start-up Leap Package" of the Korea Startup Promotion Agency under the Ministry of Small and Medium-Sized Enterprises, which supports companies in the 3rd to 7th year of startups	5 question	[5–7]
	Human resource	Creativity needs to be nurtured through education, but Korean education is thoroughly education focused on entrance examination and memorization. There is an extremely shortage of talents who are progressively challenging	5 question	
	Finance	When examining the development stage of a company on the basis of financing, the startup stage needs initial funding, the growth stage requires funding for commercialization, and the maturity stage requires funding for sustained growth	5 question	
Win-win platforms	Win-win platforms	The "win-win platform" is a place configured to exchange ideas, knowledge, information, and opinions necessary for pursuing win-win. It transforms the existing vertical cooperation network into a horizontal structure system for collaboration and coexistence between participants and between participants. It means an innovation platform that promotes the growth of the entire platform by establishing an operating body that coordinates and manages autonomous cooperation	5 question	[8]
Overcoming death-valley	Overcoming death-valley	It is common for startups to experience a shortage of funds in the process of developing the original technology and then proceeding to the proof of concept stage, and mid-sized companies succeed in developing prototypes, but fund them in the stage of implementing products that are marketable. Difficult to secure	5 question	[9]

3 Results and Discussion

3.1 *Empirical Discussion*

3.1.1 Analysis of Demographic Characteristics and Descriptive Statistics

The total number of samples used in this study was 264; frequency analysis was performed to determine the demographic characteristics of the sample. The analysis showed 80.7% (213) for men and 19.3% (51%) for women. Age was the highest in the 30s to 38.3% (101), followed by the 40s which was 34.5% (91). The startup period was the highest in less than 4–5 years at 29.9% (79), and the experience of closing was 80.7% (213). The experience of re-establishment was 53.8% (142) once and twice 29.2% (77). In this study, data collected from actual startup entrepreneurs were analyzed, and individual measurement variables were found to follow the normal distribution, as the standard deviation was less than 3, the skewness statistic was less than 3, and the kurtosis statistic was less than 3.

3.1.2 Exploring Factors and Reliability Analysis

Perform exploratory factorial analysis for validation. Principal component analysis was used, and direct church pre-use (varimax) was used to simplify the factor loading value. The selection criteria of the question in this study eigenvalue were based on 1.0 or more; factor loading is 0.4 or more. As a result of the analysis of factors, human resources 4 and 5 were removed because the construct validity was inhibited; government policy 5 was removed by the convergent validity inhibited. Exploratory factor analysis and reliability analysis summary is shown in the following (Table 2). The reliability analysis is determined that Cronbach's alpha coefficient of all measurement variables appears to be above the general level of 0.6 or higher, ensuring the reliability. The reliability analysis and exploratory factor analysis summary of the results is the following (Table 2).

3.1.3 Pearson's Correlation Analysis

Correlation analysis results showed that the correlation between variables is statistically significant. Pearson's correlation was highest with a 0.714** correlation between the win-win platform and overcoming death calley. The summary of the correlation analysis is as follows (Table 3):

Table 2 Summary of exploring factors and reliability analysis results

	Exploratory factor analysis						Reliability analysis
Item	Policy	Human resource	Finance	Win-win platforms	Overcoming death-valley	commonality	Cronbach's α
Finance3	0.787	0.166	0.233	0.211	0.104	0.708	0.880
Finance2	0.770	0.195	0.133	0.102	0.159	0.565	
Finance5	0.759	0.244	0.155	0.123	0.187	0.730	
Finance4	0.735	0.194	0.251	0.145	0.111	0.662	
Finance1	0.665	0.160	0.312	0.218	−0.013	0.667	
Overcoming death-valley2	0.192	0.769	0.233	0.259	0.073	0.623	0.895
Overcoming death-valley4	0.255	0.761	0.265	0.265	0.121	0.520	
Overcoming death-valley3	0.291	0.720	0.223	0.160	0.141	0.624	
Overcoming death-valley5	0.219	0.715	0.281	0.203	0.143	0.679	
Overcoming death-valley1	0.133	0.688	0.243	0.159	0.170	0.747	
Win-win platforms2	0.255	0.260	0.757	0.172	0.154	0.677	0.888
Win-win platforms3	0.207	0.301	0.741	0.264	0.112	0.698	
Win-win platforms4	0.232	0.283	0.708	0.190	0.156	0.581	
Win-win platforms5	0.335	0.287	0.626	0.150	0.228	0.773	
Win-win platforms1	0.322	0.282	0.588	0.229	0.132	0.758	
Policy1	0.189	0.155	0.118	0.788	0.119	0.686	0.812
Policy3	0.266	0.194	0.040	0.783	0.088	0.656	
Policy4	0.189	0.239	0.308	0.646	0.122	0.614	
Policy5	0.201	0.287	0.370	0.634	0.099	0.758	
Human resource1	0.102	0.116	0.103	0.073	0.790	0.698	0.639
Human resource2	0.114	0.106	0.258	0.214	0.698	0.802	
Human resource3	0.323	0.255	0.074	0.240	0.523	0.706	
Kaiser-Meyer-Olkin sample fit						0.931	
Bartlett's unit matrix test		Approximate chi-square	3402.942	df	231	*p* value	0.000

Table 3 Summary of correlation analysis results

Constructs	N	Mean	Standard deviation	Policy	Human resource	Finance	Win-win platforms	Overcoming death-valley
Policy	264	3.972	0.621	1	0.506[a]	0.483[a]	0.585[a]	0.567[a]
Human resource	264	3.637	0.680	0.506[a]	1	0.449[a]	0.530[a]	0.507[a]
Finance	264	3.595	0.623	0.483[a]	0.449[a]	1	0.648[a]	0.588[a]
Win-win platforms	264	3.855	0.633	0.585[a]	0.530[a]	0.648[a]	1	0.714[a]
Overcoming death-valley	264	3.799	0.653	0.567[a]	0.507[a]	0.588[a]	0.714[a]	1

[a]The correlation is significant at the 0.01 level (both sides)

3.1.4 Hypothesis Test Result

In order to test this study, the hypothesis was analyzed using SPSS ver.22 statistical package, Durbin-Watson determined that there are no residuals close to 2.105, and VIF (variance inflation factor) is less than 10 because it is determined that there is no multicollinearity between independent variables. It has been shown that the components of the entrepreneurial ecosystem have a political impact on overcoming death-valley. The hypotheses H1-1, H1-2, and H1-3 have all been adopted, and financial support has the greatest impact at 0.357***, and the next government policy appears to have an impact of 0.295***. The component of the entrepreneurial ecosystem can be said to have a 47.7% explanation for overcoming death-valley. The following (Table 4) summarizes the results of regression analysis between the entrepreneurial ecosystem components and the overcoming of death-valley, the entrepreneurial ecosystem components and the win-win platform, the win-win platform, and the death-valley overcoming.

3.1.5 Mediated Effect Analysis

Multiple regression analysis was performed to analyze the mediating effect using the three-step approach of Baron & Kenny [11] to analyze the relationship of influence of the components of the startup ecosystem on overcoming Death-Valley. In step 1, the startup ecosystem components appeared to have an effect of the (+) in overcoming death-valley hypothesis, and H1-1, 2, 3 was adopted; the startup ecosystem components in step 2 appeared to have an effect of the (+) on the win-win platform, and hypothesis H2-1, 2, 3 was adopted. In phase 3, government policy, human resources, and financial support were partially mediated in relation to the impact of the entrepreneurial ecosystem components on overcoming death-valley, and the hypothesis H4-1, 2, 3 was all adopted. Figure 2 analyzes the impact of the relationship between the components of the entrepreneurial ecosystem and the win-win platform on overcoming Death-Valley.

Table 4 Summary of multi-regression analysis results

Hypothesis	Constructs	*B*	*β*	*t*	*p* value	VIF	Results
	(Constant)	0.503		2.329	0.021		
H1-1	Policy	0.305	0.295	5.350	0.000	1.512	Accept
H1-2	Human resource	0.200	0.198	3.659	0.000	1.451	Accept
H1-3	Finance	0.381	0.357	6.704	0.000	1.407	Accept
R2 = 0.477, adjusted R2 = 0.471, $F = 79.107(p = <0.001)$, Durbin-Watson = 2.105, dependent variable: Overcoming death-valley							
	(Constant)	0.551		2.873	0.004		
H2–1	Policy	0.274	0.280	5.432	0.000	1.512	Accept
H2–2	Human resource	0.190	0.198	3.924	0.000	1.451	Accept
H2–3	Finance	0.431	0.424	8.539	0.000	1.407	Accept
R2 = 0.544, adjusted R2 = 0.539, $F = 103.296(p = <0.001)$, Durbin-Watson = 2.117, dependent variable: Win-win platforms							
	(Constant)	0.887		5.911	0.000		
H3	Win-win platforms	0.752	0.714	16.503	0.000		Accept
R2 = 0.510, adjusted R2 = 0.508, $F = 272.365(p = <0.001)$, Durbin-Watson = 2.064, dependent variable: overcoming death-valley							

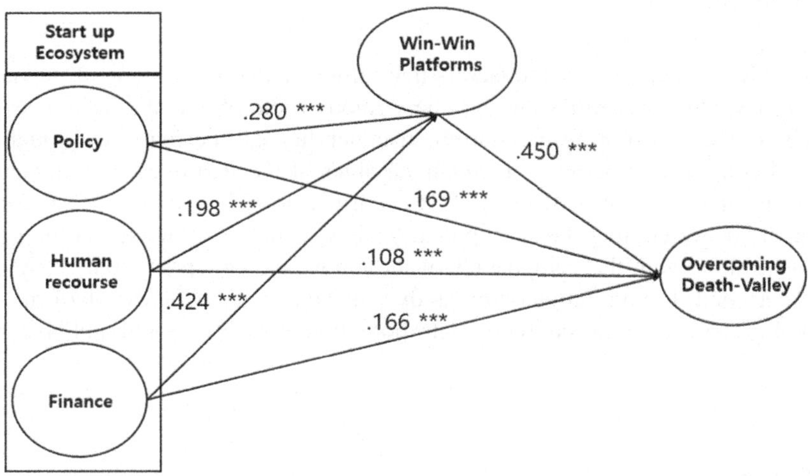

Fig. 2 Analysis of the effect of the relationship between the founding ecosystem components and the win-win platform on overcoming death-valley

Table 5 summarizes the results of the analysis of the impact of relationship between components of the startup ecosystem and the win-win platform on overcoming death-valley.

Table 5 Summary of results of mediating effects of win-win platform between startup ecosystem and overcoming death-valley

Model	Step1 (Dependent variable: overcoming death-valley)		Step2 (Dependent variable: win-win platforms)		Step3 (Dependent variable: overcoming death-valley)		Results
	B	Beta	*B*	Beta	*B*	Beta	Mediating effect
Constructs	0.503		0.497		0.242		
Policy	0.305	**0.295*****	0.327	**0.280*****	0.175	**0.169*****	Partial mediating
Human resource	0.200	**0.198*****	0.159	**0.198*****	0.110	**0.108*****	Partial mediating
Finance	0.381	**0.357*****	0.412	**0.424*****	0.177	**0.166*****	Partial mediating
Win-win platforms					0.474	**0.450*****	
R2	**0.477**		**0.544**		**0.510**		
F	**79.107*****		**103.296*****		**272.365*****		
Durbin-Watson	2.105		2.117		2.064		
p value: *$p < 0.05$ **$p < 0.01$ ***$p < 0.001$							

4 Conclusion

This study aims to find out the factors that influence the overcoming of the death-valley of startup companies through the virtuous cycle of Korea's startup ecosystem. It has demonstrated that startup sustainability can be realized through the optimal combination between the main variables of the startup ecosystem components and the win-win platform [12–16]. Furthermore, it is believed that further research on overcoming the startup death-valley, which is immersed in the fourth industry, along with the components of the innovative entrepreneurial ecosystem, will be needed. The limitation of this study is that the area of win-win platforms was limited to companies in the Seoul metropolitan area such as Seoul and Gyeonggi Province.

References

1. Hwang, S., Kim, K., Park, H.: The success factors of domestic start-up companies: case of Woowa brothers. J. Kor. Ind. Inform. Syst. Res. **25**(1), 71–87 (2020)
2. Fraiberg, S.: Start-up nation: studying transnational entrepreneurial practices in Israel's start-up ecosystem. J. Bus. Tech. Commun. **31**(3), 350–388 (2017). https://doi.org/10.1177/1050651917695541

3. Cho, Y.J., Rho, S.M., Kim, M.: Startups location analysis system using the public data. J. Digit. Contents Soc. **19**(3), 425–433 (2018)
4. Kwon, B.R., Kim, J.S.: A study on the strategy for innovation platform management to promote venture ecosystem. Electron. Telecommun. Trends. **29**(4), 137–146 (2014)
5. Sim, Y., Gil, W., Kim, S.: Study on the perception gap between technology based entrepreneurs and startup facilitators about success factors for startup. Asia Pac. J. Bus. Ventur. Entrep. **10**(2), 43–57 (2015)
6. Kim, S., Kim, M.S.: A study on the effects of regional context on entrepreneurial orientation. J. Korean Soc. Qual. Manag. **47**(4), 847–859 (2019)
7. Yoo, S., Choi, K.: How to inflow the fund for initial start-up companies using the on-line clustering platform. J. Inst. Internet, Broadcast. Commun. **15**(3), 181–189 (2015). https://doi.org/10.7236/JIIBC.2015.15.3.181
8. Mishra, S., Tripathi, A.R.: Platform business model on state-of-the-art business learning use case. Int. J. Financ. Eng. **07**, 2050015 (2020). https://doi.org/10.1142/S2424786320500152
9. Corallo, A., Errico, F., Latino, M.E., Menegoli, M.: Dynamic business models: a proposed framework to overcome the death-valley. J. Knowl. Econ. **10**(3), 1248–1271 (2019)
10. Han, J.H., Chung, J.Y.: New firm of Korean movie industry, next entertainment world's strategy. J. Korea Contents Assoc. **15**(6), 62–72 (2015). https://doi.org/10.5392/JKCA.2015.15.06.062
11. Baron, R.M., Kenny, D.A.: The moderator–mediator variable distinction in social psychological research: conceptual, strategic, and statistical considerations. J. Pers. Soc. Psychol. **51**(6), 1173 (1986)
12. Mishra, S., Tripathy, H.K., Mallick, P.K., Bhoi, A.K., Barsocchi, P.: EAGA-MLP—an enhanced and adaptive hybrid classification model for diabetes diagnosis. Sensors. **20**(14), 4036 (2020)
13. Mishra, S., Mallick, P.K., Jena, L., Chae, G.S.: Optimization of skewed data using sampling-based preprocessing approach. Front. Public Health. **8**, 274 (2020)
14. Mishra, S., Tripathy, H.K., Mishra, B.K.: Implementation of biologically motivated optimisation approach for tumour categorisation. Int. J. Comput. Aided Eng. Technol. **10**(3), 244–256 (2018)
15. Mishra, S., Mallick, P.K., Tripathy, H.K., Bhoi, A.K., González-Briones, A.: Performance evaluation of a proposed machine learning model for chronic disease datasets using an integrated attribute evaluator and an improved decision tree classifier. Appl. Sci. **10**(22), 8137 (2020)
16. Mallick, P.K., Mishra, S., Chae, G.S.: Digital media news categorization using Bernoulli document model for web content convergence. Pers. Ubiquit. Comput. (2020). https://doi.org/10.1007/s00779-020-01461-9

A Study on the Effect of the Collaboration Motive of Partners Participating in the Consulting Project on Relationship Commitment and Collaboration Performance

Chae-Sang Shin, Yen-Yoo You, and In-Chae Park

1 Introduction

Recently, SMB (small and medium business) is complaining about difficulties for survival and sustainable management in an environment of uncertainty about the future and fierce competition due to the shift from producer-centered to a customer-centric paradigm. Collaboration between companies in the supply chain is being emphasized to effectively respond to changes in the business environment amid diversified and rapid changes in customer demands [1]. Management consulting is a representative knowledge service industry that is knowledge-intensive and creates added value. The role of the knowledge-intensive service industry is inputted as intermediate goods in a series of processes from production planning to delivery of value to customers through the distribution process in the overall value chain of the industry. It plays a role as a catalyst for promoting industrial innovation and plays a fundamental role in innovation as a performer of innovation [2]. SMB can identify internal problems that they did not recognize through management consulting and derive plans for sustainable growth. Consulting as an external expert contributes to enhancing the competitiveness of a company by improving competitiveness at the corporate, industrial, and national levels and by supporting problem-solving or management environment improvement. The consulting project is part of the knowledge service industry that provides solutions to various problems faced by SMB based on their knowledge and experience in a network process system where consultants from various fields participate [3]. Therefore, the purpose of this study is to

C.-S. Shin (✉)
Department of Smart Convergence Consulting, Hansung University, Seoul, South Korea

Y.-Y. You · I.-C. Park
Division of Smart Management Engineering, Hansung University, Seoul, South Korea
e-mail: threey0818@hansung.ac.kr; ipark@hansung.ac.kr

S. R. Samanta et al. (eds.), *Cognitive Computing for Risk Management*,
EAI/Springer Innovations in Communication and Computing,
https://doi.org/10.1007/978-3-030-74517-2_5

empirically analyze the causal relationship and the mediating role of relationship commitment that a collaboration motive of the partner participating in the consulting project affects the collaboration performance.

2 Materials and Methods

2.1 Literature Review

2.1.1 Management Consulting

Management consulting is defined as independent and professional advising services that help the management organization achieve the goals pursued by solving corporate problems and discovering, utilizing, and applying new opportunities [4]. The basic structure of management consulting begins with a single relationship between a consultant and a client but is potentially formed by a combination of various factors [5]. If the consultant predicts the market situation in the wrong direction due to the lack of expertise, or if inefficiency occurs due to the consultant's negligence and low commitment, successful consulting execution and completion cannot be achieved [6]. The effectiveness of consulting can be influenced by the characteristics of the company. In particular, since the effect may vary depending on the size of the company, industry field, and innovation structure, it is required to establish a consulting strategy appropriate for it [7]. As the executioner of actual consulting for successful management consulting, consultant collaboration competency, relationship trust, and relationship commitment to achieve the purpose of consulting execution are directly related to consulting performance.

2.1.2 Consultant's Collaboration Motive

Collaboration is driven by the diversification effect of risk among collaborative partners [8]. The risk of a large-scale project with uncertain performance can be distributed to one or more companies, and it is promoted to reduce the risk in the market by enabling market diversification and to achieve cost reduction effects [9]. In the case of companies that reduce production costs or costs through collaboration and are greatly affected by the learning effect of execution, they form a cooperative relationship by performing effective and economical management activities by using the comparative advantage of the cooperative partner. It is executed to obtain a synergy effect by transferring and supplementing the skills or other skilled techniques possessed by partners [10]. By establishing a cooperative relationship through collaboration, it is possible to reduce the cost of searching for partners that can share mutually heterogeneous and exclusive resources. Deci and Ryan's self-determination theory (SDT) among motivational theories related to the collaboration motivations of the partners participating in the consulting project emphasizes the intrinsic motivation of people's actions. Humans basically have three basic psychological needs, autonomy, competence, and relevance, and emphasize that this is not a need formed through learning but a need for a unique human nature that transcends sex, culture, and time [11]. The three basic psychological needs of humans are as follows. First, autonomy is the desire to feel one's choice and will in action, wanting to be the subject of an individual's actions, and wishing for the freedom to decide what is important and valuable to the individual [12]. When a behavior is autonomous, our interests, preferences, and needs guide our decision-making process about whether or not to engage in a particular activity. Second, competence reflects the desire to cultivate individual competencies and skills throughout life and to pursue and master the optimal challenge in doing so [13]. It is also the desire to feel effective in action and achieve the result it is intended to achieve. Competency is not the acquired skill itself but the desire to feel confidence and efficiency through the mind and actions that want to perceive one's competence [14]. Third, relevance is trying to experience a sense of belonging to be with others in many activities in life [11]. This reflects the desire to be emotionally connected or interrelated in a warm relationship. In a relationship that satisfies the need for relevance, what people essentially seek is an opportunity for themselves and others to have a real relationship in an emotionally meaningful way [11]. Collaboration can secure competitiveness in the market and increase synergy by sharing complementary core competencies. In addition, collaboration is an action in which stakeholders achieve common goals and create results through communication and cooperation, and jointly promote projects by linking individual specialized functions.

2.1.3 Relationship Commitment

Commitment is a social construct in which people interact to achieve the common goals of an organization. Prior research has been conducted from the three viewpoints of the relationship commitment of the consulting project participants. First, effective commitment means emotional attachment and unity that includes positive efforts to accept and meet the needs of consulting companies. Second, continuance commitment means that the cost incurred by leaving the organization is becoming more and more immersed to remain in the organization, which means that it can lead effective consulting by sustaining a long-term relationship with the recipient company. Third, normative commitment means a commitment to the organization with a sense of psychological or moral obligations and values [15]. Relationship commitment is an important variable that separates relationship breakers and relationship persisters and is defined as a persistent desire to maintain a valuable relationship. Therefore, relationship commitment implies the importance of the relationship with the partner and the desire to continue the relationship until the future. Commitment in a relationship is a perception that the relationship will bring future value or benefit to partners [16], while at the same time a belief that a lasting relationship with others deserves the greatest effort in maintaining it [17], and intent to make short-term sacrifices to realize long-term benefits [18]. Relationship commitment exists only when relationships are considered important to the parties and the willingness to strive to maintain them is felt [17]. In addition, Berry and Parasuraman [19] argued that relationships are formed on the basis of mutual commitment and that the engaged parties deserve to ensure that the relationship continues indefinitely. In the interorganizational relationship, on the other hand, real cooperation can be promoted only when all participants participate in advance, and a high level of relationship commitment provides a situation in which the two partners can achieve their personal and common goals. Therefore, immersive partners will do their best to each other's corporate relationship, and since they solve short-term problems in terms of achieving long-term goals, high-level relationship commitment is expected to have a very close relationship with the success of value chain partnerships.

2.1.4 Collaboration Performance

In order to expect the positive effects of collaborative activities, there is a task to resolve various side effects such as conflict within the team. Since each team has different methods and processes for performing tasks, and factors that enable effective interactions may differ depending on the nature of the team, it is necessary to analyze the factors that affect the team's performance [20]. The source of collaboration is the shared goals of members for consistency and direction of collaboration, clear responsibilities, and roles for each member's task through determining the quality and quantity of tasks, and motivation through providing fair awareness of the outcome of the task. It is a leading example based on the mutual recognition and action of each member that can influence the rewards and balances and the creation of results [20]. Collaboration performance means transaction cost or the cost and efficiency of relationship management, and in some cases, it means the degree of effectiveness [21]. There is a trap that it is difficult to expect collaboration that generates results that occur in the process of collaboration. "Collaboration opportunity was evaluated to overcome four pitfalls such as straps from silo culture," "excessive collaboration," "overevaluation of collaborative value," and "underestimation of collaborative cost," and to achieve systematic collaboration. It is necessary to identify barriers to collaboration and implement customized solutions [22]. Regarding collaboration performance, project performance is the result of collaboration activities, and the emphasis will be different depending on the relationship capabilities of partners, collaboration motive, relationship trust, and relationship commitment. The result of the collaboration will contribute to productivity and profitability, growth, and stability, and to the spread of innovation among project partners.

2.2 *The Research Model and Hypothesis*

2.2.1 Research Model

This study analyzes the causal relationship through regression analysis between the collaboration motive and collaboration performance of partners participating in SMB's consulting projects in the B2B knowledge service industry. Therefore, the purpose of this study is to analyze the mediating effect of relationship commitment in the relationship between the collaboration motive and collaboration performance. Based on the theoretical considerations of previous studies, the collaboration motive was classified into autonomy, competence, and relevance, and selected as an independent variable. The dependent variable was selected as collaboration performance. Relationship commitment was selected as a parameter in the relationship between collaboration motive and collaboration performance. Based on these preceding studies, a research hypothesis was established to empirically analyze the impact relationship to increase the collaboration performance of consulting projects in the knowledge service industry. The research model for the empirical analysis of this study is as follows (Fig. 1).

2.2.2 Research Hypothesis

Based on the theoretical considerations of the previous research, the research hypothesis was established to test the causal relationship between the selected independent variables, the collaboration motive, and the dependent variables, the collaboration performance. The hypothesis on the mediated effect of relationship commitment in the causal relationship between collaboration motive and collaboration performance is set as follows:

H1. Autonomy will have a positive effect on collaboration performance.
H2. Competence will have a positive effect on collaboration performance.
H3. Relevance will have a positive effect on collaboration performance.
H4. Autonomy will have a positive effect on relationship commitment.

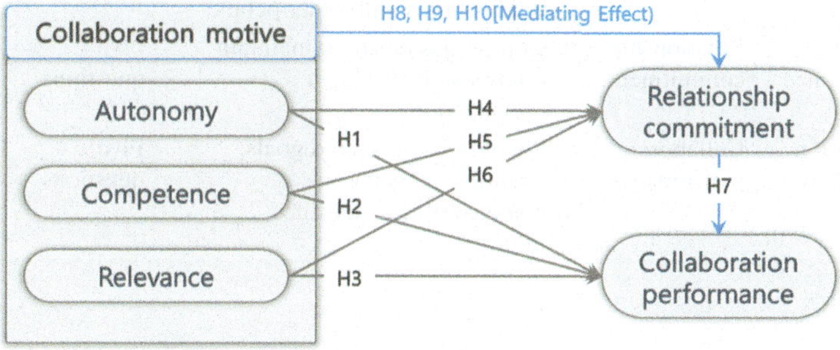

Fig. 1 Research model

H5. Competence will have a positive effect on relationship commitment.
H6. Relevance will have a positive effect on relationship commitment.
H7. Relationship commitment will have a positive effect on collaboration performance.
H8. Relationship commitment will mediate the effect of autonomy on collaboration performance.
H9. Relationship commitment will mediate the effect of competence on collaboration performance.
H10. Relationship commitment will mediate the effect of relevance on collaboration performance.

2.2.3 Operational Definition of Variables

Based on the previous research, this study classified the motivation of collaboration of consulting partners into autonomy, competence, and relevance, and is composed of a total of 15 questionnaires, each with 5 questions. Next, the relationship commitment was composed of five questionnaires and five questionnaires for collaboration performance. It is organized as shown in Table 1 in relation to the operational definition of measurement variables for collaboration motive, relationship commitment, and collaboration performance.

Table 1 Summary of operational definitions of measured variable

Evaluation items	Measurement variable	Operational definition	Questions	Researcher
Collaboration motive	Autonomy	Trying to control and decide one's own behavior, free from external control	Five questions	[8–14]
	Competence	To experience the experience or efficacy of challenge and skill	Five questions	
	Relevance	Trying to have an active interaction with each other by maintaining intimacy in relationships with other people	Five questions	
Relationship commitment	Relationship commitment	Continuous desire to maintain relationships between organizations	Five questions	[15–19]
Collaboration performance	Collaboration performance	Shared vision, shared goals, rewards and balances, responsibilities and roles, initiatives	Five questions	[20–22]

3 Results and Discussion

3.1 Empirical Analysis Results

3.1.1 Demographic Characteristic Analysis

In this study, a total of 352 data were collected using a questionnaire for consultants currently conducting consulting. However, the number of samples used for statistical analysis was 323, excluding 29 unfaithful questionnaires. Statistical analysis for data analysis utilized the SPSS ver.22 statistical package. Frequency analysis was conducted to grasp the distribution of the general characteristics of the collected samples. As a result of frequency analysis, males showed a high rate of 73% (236) and females 27% (87). As for the age distribution, the 50s showed the highest at 37.2% (120), followed by the 40s at 31% (100), showing the distribution of experienced consultants. In terms of education, university graduation was the highest with 47.0% (152), followed by graduate degree or higher with 36.5% (118). It is analyzed that the consultant's high academic background has the knowledge necessary for consulting. Consulting experience was evenly distributed. Those over 5 years and under 7 years were the highest at 24.1% (78), followed by those over 9 years in the order of 21% (68). Through frequency analysis, it can be guessed that the years of experience (knowledge and experience) of consultants participating in consulting projects are high.

3.1.2 Exploratory Factor Analysis and Reliability Analysis

For the analysis of validity and reliability, data is collected by constructing a questionnaire based on previous studies. Validity and reliability are secured through the composition of the questionnaire and exploratory factor analysis and reliability analysis of the collected data. Exploratory factor analysis and reliability analysis were performed to verify the validity and reliability of the measurement items. In exploratory factor analysis, the principal component analysis was selected as the factor extraction method, and Varimax with Kaiser normalization was selected as the rotation method. The eigenvalue was set to be 1.0 or more as a selection criterion for measurement items. In this study, exploratory factor analysis was performed five times. Finally, the KMO (sample suitability) index value is 0.920, which is above the general level of 0.7. As a result of Bartlett's unit matrix test, the $x^2(p)$ value was 0.000, which was lower than the general level of 0.05, making it suitable for exploratory factor analysis [23]. Through factor analysis, competency 5 was the first to be eliminated as it inhibited discriminant validity among the measurement items that hindered concentration validity and discriminant validity. Next, they were removed in the order of relationship commitment 1, Competency 4, and Collaboration performance 2, and classified into 5 factors. Reliability analysis for reliability verification was refined in exploratory factor analysis, and internal consistency was analyzed using Cronbach's alpha coefficient to secure homogeneity between measurement items that secured single dimensionality. As a result of reliability analysis, Cronbach's alpha coefficient of all measurement items was found to be above the general level of 0.6. It is judged that reliability is secured even without removing separate measurement items [23]. The summary of the exploratory factor analysis and reliability analysis results of the measurement items for the questionnaire is shown in Table 2.

Table 2 Summary of validity and reliability analysis

Item	Exploratory factor analysis						Reliability
	Rela-Comm	Rele	Auto	Coll-Perf	Comp	Comm	Cronbach's alpha
Rela-Comm 4	**0.725**	0.271	0.142	0.290	0.113	**0.717**	**0.881**
Rela-Comm 5	**0.725**	0.229	0.114	0.284	0.142	**0.692**	
Rela-Comm 2	**0.679**	0.253	0.142	0.321	0.170	**0.678**	
Rela-Comm 3	**0.615**	0.373	0.276	0.242	0.155	**0.677**	
Rele 2	0.026	**0.705**	−0.002	0.290	0.224	**0.631**	**0.853**
Rele 5	0.399	**0.685**	0.067	0.303	0.021	**0.726**	
Rele 4	0.270	**0.682**	0.023	0.306	0.110	**0.645**	
Rele 1	0.434	**0.681**	0.169	0.029	0.065	**0.686**	
Rele 3	0.380	**0.669**	0.232	0.084	0.091	**0.661**	
Auto 5	−0.059	−0.024	**0.853**	0.037	0.026	**0.734**	**0.832**
Auto 4	0.133	0.047	**0.839**	0.169	0.068	**0.757**	
Auto 2	0.349	0.048	**0.702**	0.170	0.090	**0.654**	
Auto 3	0.063	0.214	**0.686**	0.142	0.164	**0.568**	
Auto1	0.382	0.199	**0.577**	0.129	0.116	**0.548**	
Coll-Perf 4	.262	0.134	0.178	**0.797**	0.132	**0.771**	**0.864**
Coll-Perf 3	0.361	0.222	0.106	**0.720**	0.090	**0.717**	
Coll-Perf 1	0.088	0.291	0.213	**0.713**	0.130	**0.663**	
Coll-Perf 5	0.405	0.232	0.188	**0.686**	0.145	**0.745**	
Comp 1	0.230	0.051	0.081	0.062	**0.836**	**0.764**	**0.655**
Comp 3	0.321	0.165	0.186	0.254	**0.609**	**0.600**	
Comp 2	−0.175	0.461	0.205	0.193	**0.586**	**0.666**	
Eigenvalue	3.309	3.211	3.151	2.939	1.690		
% of variance	15.755	15.293	15.006	13.994	8.049		
% of accumulated	15.755	31.048	46.054	60.048	**68.096**		
Kaiser-Meyer-Olkin sample fit							**0.920**
Bartlett's unit matrix test		Chi square	3748.291	df	210	*p*-**value**	**0.000**

Rela-Comm relationship commitment, *Rele* relevance, *Auto* autonomy, *Coll-Perf* collaboration performance, *Comp* competence, *Comm* commonality

Table 3 Summary of correlation analysis results

Constructs	*N*	Mean	Standard deviation	Auto	Comp	Rele	Rela-Comm	Coll-Perf
Auto	323	3.24	0.650	1	0.396[a]	0.362[a]	0.455[a]	0.446[a]
Comp	323	3.64	0.654	0.396[a]	1	0.500[a]	0.475[a]	0.487[a]
Rele	323	3.97	0.603	0.362[a]	0.500[a]	1	0.688[a]	0.612[a]
Rela-Comm	323	3.96	0.644	0.455[a]	0.475[a]	0.688[a]	1	0.678[a]
Coll-Perf	323	3.76	0.660	0.446[a]	0.487[a]	0.612[a]	0.678[a]	1

Coll-Perf collaboration performance, *Comp* competence, *Rela-Comm* relationship commitment, *Rele* relevance, *Auto* autonomy

[a]The correlation is significant at the 0.01 level (both sides)

3.1.3 Correlation Analysis

In this study, through exploratory factor analysis, measurement items that hindered validity were removed, and single dimensionality was secured. Reliability was secured through reliability analysis using Cronbach's alpha coefficient to ensure consistency between the items of measurement and to verify that there is no problem even when creating a variable by taking an average. The correlation coefficient (r) has a value of −1 to 1, and if it is close to "−1," it is a negative (−) correlation, if it is "0," there is no correlation, and if it is close to "1," it is positive (+) correlation. It was newly composed of five measurement variables, autonomy, competence, relevance, relationship commitment, and collaboration performance, which were variables calculated through averaging among the measurement items whose validity and reliability were confirmed. Correlation analysis was performed to find out the direction and density between newly constructed measurement variables. Looking at the correlation between the measured variables in Table 3, autonomy showed a high correlation with relationship commitment as 0.455**, followed by collaboration performance and Pearson correlation coefficient as 0.446**. Relevance shows high correlation with relationship commitment as 0.688**, and collaboration performance and high correlation with 0.612**. The correlation between relationship commitment and collaboration performance was high as 0.678**. When looking at the results of the correlation analysis between all the measured variables, all statistically significant positive correlations were shown. It is judged that regression analysis is possible to grasp the causal relationship between constructs. The results of the correlation analysis are summarized as follows (Table 3).

3.1.4 Hypothesis Test Result

To test the hypothesis of this study, multiple regression analysis was performed using the SPSS ver.22 statistical package. In Table 4, the Durbin-Watson index was 2.117 in the causal relationship between the collaboration motive and the collaboration performance, 2.089 in the causal relationship between the collaboration motive and the relationship commitment, and 2.008 in the causal relationship between the relationship commitment and collaboration performance. The Durbin-Watson index

Table 4 Summary of multiple regression analysis result

Hypothesis	Constructs	*B*	*β*	*t*	*p*-value	VIF	Results
	(Constant)	0.457		2.235	0.026		
H1	Autonomy	0.217	0.214	4.641	0.000	1.238	accept
H2	Competence	0.181	0.180	3.630	0.000	1.435	accept
H3	Relevance	0.487	0.445	9.114	0.000	1.393	accept
Durbin-Watson = 2.117, $R^2 = 0.455$, adjusted $R^2 = 0.450$, $F = 88.674$ ($p = <0.001$),							
Dependent variable, collaboration performance							
	(Constant)	0.523		2.828	0.005		
H4	Autonomy	0.207	0.209	4.896	0.000	1.238	accept
H5	Competence	0.114	0.115	2.512	0.012	1.435	accept
H6	Relevance	0.593	0.555	12.275	0.000	1.393	accept
Durbin-Watson = 2.089, $R^2 = 0.532$, adjusted $R^2 = 0.527$, $F = 120.772$ ($p = <0.001$), Dependent variable, relationship commitment							
	(Constant)	1.002		5.936	0.000		
H7	Relationship commitment	0.695	0.678	16.522	0.000		accept
Durbin-Watson = 2.008, $R^2 = 0.460$, adjusted $R^2 = 0.458$, $F = 272.964$ ($p = <0.001$), Dependent variable, collaboration performance							

is to secure independence between residuals (meaning autocorrelation of the residual). A value between 0 and 4 and close to 2 means no autocorrelation. In this study, independence between residuals was secured. To diagnose the multicollinearity between the independent variables, when looking at the index of variance expansion coefficient (VIF), all values were found to be less than 10, and the correlation between the independent variables was analyzed to below. Therefore, it is judged to be suitable for multiple regression analyses because the independence of independent variables is secured. Looking at the non-standardization coefficient (*B*) in the causal relationship between the collaboration motive and the collaboration performance of the consulting project participants, it was found that the collaboration motive variables such as autonomy, competence, and relevance all had a positive (+) effect on the collaboration performance. The hypotheses H1, H2, and H3 were all adopted. Next, looking at the standardization factor (*β*), it was analyzed that relevance showed the highest effect relationship on collaboration performance at 0.445***. The explanatory power (R^2) for the relationship between the collaboration motive and the collaboration performance was high at 0.455 (46%). Next, looking at the non-standardization coefficient (*B*) in the causal relationship between the collaboration motive and relationship commitment of the consulting project participants, it was found that the collaboration motive had a positive (+) effect on the relationship commitment. The hypotheses H4, H5, and H6 were all adopted. Next, looking at the standardization factor (*β*), it was analyzed that relevance has the highest effect on relationship commitment at 0.555***. The explanatory power (R^2) of the impact relationship between the collaboration motive and relationship commitment was high at 0.532 (53%). Lastly, looking at the non-standardization coefficient (*B*) in the causal relationship between relationship commitment and collaboration motive, it was found that relationship commitment had a positive (+) effect on collaboration performance. The hypothesis of hypothesis H7 was adopted. It can be said that the explanatory power (R^2) of the relationship between relationship commitment and collaboration performance is 0.460 (46%). The following (Table 4) summarizes the results of the hypothesis test between collaboration motive and collaboration performance, hypothesis test between collaboration motive and relationship commitment, and hypothesis test between relationship commitment and collaboration performance of consulting project participants through multiple regression analysis.

3.1.5 Mediated Effect Analysis

To analyze the mediating effect of relationship commitment in the causal relationship of collaboration motives of consulting project participants on collaboration performance, multiple regression analysis was performed using the three-step approach of Baron and Kenny [24]. First, in Step 3, it was analyzed that all of the relationship commitment partially mediated the causal relationship that the collaboration motive of the consulting project participants had on the collaboration performance. Table 5 shows a summary of the results of the mediating effect analysis of relationship commitment in the causal relationship between the collaboration motive and collaboration performance of consulting project participants. H7's hypothesis that "Relationship commitment will have a positive (+) effect on Collaboration performance" was adopted. H8, H9, and H10 were all adopted in the hypothesis test for the mediating effect. In the results of the relationship commitment, which was put as a mediating in the causal relationship between collaboration motives and collaboration performance, the standardization coefficient was found to be statistically significant as 0.404***(Beta). Relationship commitment accounts for 0.531 (53%) of explanatory power (R^2) for the causal relationship of collaboration performance. Comparing Step 1 and Step 3 about the mediating effect analysis shows the change in the standardization coefficient (β) for the mediating effect of relationship commitment in the causal relationship that the collaboration motive has on the collaboration performance. Due to the mediating effect of relationship commitment, the standardization coefficient (β) is autonomy (0.214, $p < 0.001$~ 0.129, $p < 0.05$) and competence (0.180, $p < 0.01$~0.133, $p < 0.05$), and the influence of relevance (0.445, $p < 0.001$~0.220, $p < 0.001$) was reduced. It shows the importance of relationship commitment in the causal relationship between the collaboration motive and collaboration performance of consulting project participants [25–29]. Table 5 is a summary of the results of the mediated effect analysis of relationship commitment. Finally, Fig. 2 shows that it has an important effect on the collaboration motive and collaboration performance of the consulting project participants, as

Table 5 Summary of mediating effects of relationship commitment in collaboration motive and collaboration performance

Model	Step 1	Step 2	Step 3	Results				
	Dependent variable: collaboration performance	Dependent variable: relationship commitment	Dependent variable: collaboration performance					
	B	Beta	*B*	Beta	*B*	Beta	Mediating effect	
(Constant)	0.457		0.523		0.240			
Autonomy	0.217	0.214***	0.207	0.209***	0.131	0.129**	Accept	Partial
Competence	0.181	0.180***	0.114	0.115**	0.134	0.133**	Accept	Partial
Relevance	0.487	0.445***	0.593	0.555***	0.241	0.220***	Accept	Partial
Relationship commitment					0.414	0.404***	Accept	Mediating
R^2 (adjusted R^2)	0.455(0.450)	0.532(0.527)	0.531(0.525)					
F-value	88.674***	120.772***	90.074***					
Durbin-Watson	2.117	2.089	2.074					

p-value: $*P < 0.05$, $**p < 0.01$, $***p < 0.001$

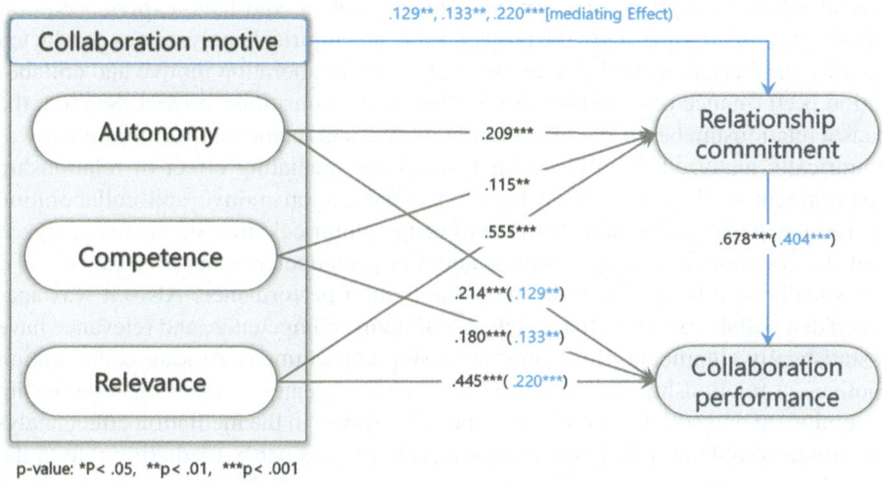

Fig. 2 Results of mediating effect of relationship commitment in collaboration motive and collaboration performance

well as relationship commitment. Besides, relationship commitment affects collaboration performance and is mediated in the causal relationship between collaboration motive and collaboration performance.

Figure 2 shows the overall results of the hypothesis test and the mediating effect of relationship commitment in the causal relationship between the collaboration motive and collaboration performance.

4 Conclusion

SMB meet diverse and complex customer needs with limited resources amid a paradigm shift from producer-centered to consumer-centered, and it is difficult to predict the uncertain future of sustainability. With the development of ICT technology, consumer demands have diversified, and their purchasing power through the community is increasing very much. To respond quickly and flexibly, companies are reorganizing the organizational structure into teams and trying to improve effective performance through cooperation. Also, we are pursuing human resources with teamwork competency and fostering a people-centered corporate culture. As a result of previous studies, most of the studies are targeting SMB that have received consulting services. SMB lack manpower and capacity to solve internal problems on their own. Therefore, consulting is requested by an external consulting company. There are many differences in collaboration performance depending on the competency, motivation, and level of commitment of the partners participating in the consulting project. This study is a study on the factors of collaboration motive and relationship commitment of project partners that affect the performance of consulting projects. It is aimed at consulting partners who have experienced collaboration by participating in the project. First, an empirical analysis was conducted on the causal relationship between the factors of collaboration motive and collaboration performance of partners participating in the consulting project. Second, the causal relationship between collaboration motive and relationship commitment was empirically analyzed. Finally, an analysis of the mediating effect of relationship commitment in the relationship between collaboration motive and collaboration performance was conducted. As a result of the empirical analysis, it was analyzed that the collaboration motive such as autonomy, competence, and relevance had a statistically significant effect on the collaboration performance. Also, it was analyzed that collaboration motives such as autonomy, competence, and relevance have a statistically significant effect on relationship commitment. Among collaboration motives, a relationship was analyzed to have the greatest impact on relationship commitment and collaboration performance. Besides, in the mediation effect analysis, it was found that relationship commitment plays a partly mediating role in the causal relationship between collaboration motive and collaboration performance. It was analyzed that the collaboration motive of the consulting partner was further enhanced through relationship commitment. The higher the collaboration motive, the higher the collaboration performance, and the higher the motivation for collaboration, the higher the relationship commitment is, emphasizing the importance of motivation and relationship commitment of partners participating in the consulting project. Therefore, this study will be important data in conducting SMB consulting with consulting companies and consultants. However, there will be many more influencing factors for collaboration performance. In future research, there is a need to expand research on the degree of cohesion and relationship trust of partners.

References

1. Jap, S.D.: Pie-expansion efforts: collaboration processes in buyer–supplier relationships. J. Market. Res. **36**(4), 461–475 (1999)
2. Hertog, P.D.: Knowledge-intensive business services as co-producers of innovation. Int. J. Innov. Manag. **4**(04), 491–528 (2000)
3. Shin, C.S., You, Y.Y.: Study on the effect of consultant's social-relational competency on trust and collaboration performance in the knowledge service industry. Int. J. Adv. Sci. Technol. **29**(4), 163–173 (2020)
4. Kubr, M. (ed.): Management Consulting: A Guide to the Profession. International Labour Organization, Geneva (2002)
5. Lee, Y.H., Seo, Y.W.: A study on the effects of business consulting on the business performance: analysis of elasticity of consulting factors and a group analysis based on firm age. Knowl. Manag. Res. **19**(4), 39–58 (2018)
6. Bae, Y.S., Ahn, Y.J.: An empirical study of the effects of consultant competency on the performance of management consulting: focusing on the moderating effects of the characteristics of consulting firm and client firm. Department of Business Administration, Graduate School Dankook University, Republic of Korea (2013)
7. Coad, A., Holm, J.R., Krafft, J., Quatraro, F.: Firm age and performance. J. Evol. Econ. **28**(1), 1–11 (2018)

8. Fuller, M.B., Porter, M.E.: Coalitions and global strategy from. Competition Glob. Ind. **315**, 344 (1986)
9. Mariti, P., Smiley, R.H.: Co-operative agreements and the organization of industry. J. Ind. Econ. **31**, 437–451 (1983)
10. Harrigan, K.R.: Exit barriers and vertical integration. Acad. Manage. J. **28**(3), 686–697 (1985)
11. Gagné, M., Deci, E.L.: Self-determination theory and work motivation. J. Organ. Behav. **26**(4), 331–362 (2005)
12. Deci, E.L., Ryan, R.M.: The "what" and "why" of goal pursuits: human needs and the self-determination of behavior. Psychol. Inq. **11**(4), 227–268 (2000)
13. Deci, E.L., Ryan, R.M.: The general causality orientations scale: self-determination in personality. J. Res. Pers. **19**(2), 109–134 (1985)
14. Ryan, R.M., Deci, E.L.: Overview of self-determination theory: An organismic-dialectical perspective. In E. L. Deci & R. M. Ryan (Eds.), Handbook of self-determination research (pp. 3–33). University of Rochester Press (2002)
15. Allen, N.J., Meyer, J.P.: The measurement and antecedents of affective, continuance and normative commitment to the organization. J. Occup. Psychol. **63**(1), 1–18 (1990)
16. Hardwick, B., Ford, D.: Industrial buyer resources and responsibilities and the buyer-seller relationships. Ind. Mark. Purch. **1**(3), 3–26 (1986)
17. Hunt, S.D., Morgan, R.M.: Organizational commitment: one of many commitments or key mediating construct? Acad. Manage. J. **37**(6), 1568–1587 (1994)
18. Dwyer, F.R., Schurr, P.H., Oh, S.: Developing buyer-seller relationships. J. Mark. **51**(2), 11–27 (1987)
19. Parasuraman, A., Berry, L.L., Zeithaml, V.A.: Understanding customer expectations of service. Sloan Manage. Rev. **32**(3), 39–48 (1991)
20. Song, J.J., Kim, M.J.: The effect of team collaboration and transactive memory system on team performance: focused on moderate effect of maturity. Gachon University, 1–119 (2013)
21. Noordewier, T.G., John, G., Nevin, J.R.: Performance outcomes of purchasing arrangements in industrial buyer-vendor relationships. J. Mark. **54**(4), 80–93 (1990)
22. Salancik, G.R., Pfeffer, J.: A social information processing approach to job attitudes and task design. Adm. Sci. Q. **23**, 224–253 (1978)
23. Choi, C.H., Roh, J.W., You, Y.Y.: A study on the effects of ties between clients and consultants on consulting project performance(focused on mediating effect of trust, moderating effect of consultant's competence and client's participation of consulting). J. Prod. Rev. **27**(3), 87–114 (2013)
24. Baron, R.M., Kenny, D.A.: The moderator–mediator variable distinction in social psychological research: conceptual, strategic, and statistical considerations. J. Pers. Soc. Psychol. **51**(6), 1173 (1986)
25. Mishra, S., Tripathy, H.K., Mallick, P.K., Bhoi, A.K., Barsocchi, P.: EAGA-MLP—an enhanced and adaptive hybrid classification model for diabetes diagnosis. Sensors. **20**(14), 4036 (2020)
26. Mishra, S., Mallick, P.K., Jena, L., Chae, G.S.: Optimization of skewed data using sampling-based preprocessing approach. Front. Public Health. **8**, 274 (2020)
27. Mishra, S., Tripathy, H.K., Mishra, B.K.: Implementation of biologically motivated optimisation approach for tumour categorisation. Int. J. Comput. Aided Eng. Technol. **10**(3), 244–256 (2018)
28. Mishra, S., Mallick, P.K., Tripathy, H.K., Bhoi, A.K., González-Briones, A.: Performance evaluation of a proposed machine learning model for chronic disease datasets using an integrated attribute evaluator and an improved decision tree classifier. Appl. Sci. **10**(22), 8137 (2020)
29. Mallick, P.K., Mishra, S., Chae, G.S.: Digital media news categorization using Bernoulli document model for web content convergence. Pers. Ubiquit. Comput. (2020). https://doi.org/10.1007/s00779-020-01461-9

A Study on the Influence of Consulting Execution Characteristics on Result Quality and Business Utilization: Focusing on Companies Participating in Government-Supported Consulting Projects

Hye-Joo Song, Yen-Yoo You, and Hyun-Sung Park

1 Introduction

As the global economy transitions to a knowledge-based economy, the consulting industry that provides solutions to corporate management problems is expected to expand further [1]. Large corporations have already established strategies to respond to the rapidly growing global economic environment [2]. Few Fortune 500 companies do not use management consultants [3]. Therefore, consulting support should be provided with the recognition that the help of external experts is needed throughout the management of SMEs that the lack of professional manpower, management funds [4]. The consulting business plays a role in improving SMEs' productivity and business management capabilities through expert diagnosis and guidance, and enhancing their internal and external competitiveness [4, 5]. The purpose of this study, in deriving and utilizing consulting results by companies that receive consulting through government-sponsored consulting programs, is to examine whether consulting performance characteristics such as the support role of the government, support capability of consultant's, and participation attitude of companies affect quality and utilization.

H.-J. Song · Y.-Y. You · H.-S. Park (✉)
Hansung University, Seoul, South Korea
e-mail: shj-629@kotra.or.kr; threey0818@hansung.ac.kr; hyunsung.park@hansung.ac.kr

S. R. Samanta et al. (eds.), *Cognitive Computing for Risk Management*,
EAI/Springer Innovations in Communication and Computing,
https://doi.org/10.1007/978-3-030-74517-2_6

2 Materials and Methods

2.1 Theoretical Background

2.1.1 Consulting Performance Characteristics

The consultant's job competency includes professional knowledge, analytical ability, strategic thinking ability, document writing ability, and communication ability [1]. Consultants must be independent in financial, decision-making, and emotive terms, so that they can express their assessments and opinions objectively and impartially [6]. Consultants often take a position to persuade others to perform various tasks with little or no authority to make a difference [7]. Trust is the intention of a party to be susceptible to the actions of counterparty based on the expectation that another party will perform a specific action against the trustee [8]. Trust means a positive expectation that others will not act opportunistically in words, actions, or decisions [9]. He suggests that managers and consultants organize participation according to their target hierarchy [10]. The alternative we have briefly outlined can reform the client-consultant relationship [3]. Clint focus, clarity of objectives, partnership with the team, and the availability of consultants during the implementation phase were both critical to improving project outcomes [11]. Appropriateness refers to the degree to which currently available knowledge or skills are used in conduction consulting [12]. The government's consulting support project provides consulting expenses to SMEs that are difficult to bear the consulting costs [4]; the use of government aid schemes depends on trust in the government [13].

2.1.2 Result Quality

Service quality is the overall impression of the service provider or the relative excellence/inferiority of its services [14]. Respondents thought that the three indicators of success in consulting were achievement of agreed goals, customer/customer satisfaction, and timeliness of service delivery [15]. The SERVQUAL model was completed by integrating the criteria for evaluating service quality into five dimensions (typicality, trustworthy, responsiveness, confidence, and sympathy) [14]. That can be seen in several researches that customer satisfaction and service quality are independent of each other [14]. The issue between satisfaction and quality relates to the causal relationship between the two concepts, the view that quality is a leading variable of satisfaction [26] and the view that satisfaction is a leading variable of quality [14, 27]. It has higher significance to service quality improvement for the sake of business success [16].

2.1.3 Business Utilization

Utilization refers to how much the results of consulting services suggested by consultants are actually utilized and applied in the workplace [17]. Innovative action refers to an activity that can improve the performance of one's task or group or organizations to which you belong [18]. Business utilization represents the degree used to perform the task [19]. Bitner's research on airport users showed the structure of "satisfaction→service quality→behavioral intention" [14]. On the contrary, Woodside et al. [26] confirmed that the structure of "service quality→satisfaction→purchase intention" is shown [14].

2.2 *Research Model*

This study aims to search whether the characteristics of consulting, such as consultant expertise, reliability of service, corporate acceptance attitude, suitability of service performance, and fidelity of support from government agencies, affect the result quality of consulting and utilization of business when small- and medium-sized enterprises who are using government-supported consulting programs are consulted. In order to do this research, the research model in Fig. 1 was derived based on the contents confirmed through previous studies. Until now, many previous studies have only verified the impact or satisfaction of consulting performance characteristics on result quality. However, this study tried to go further from here to check whether the consulting results were actually used by companies that

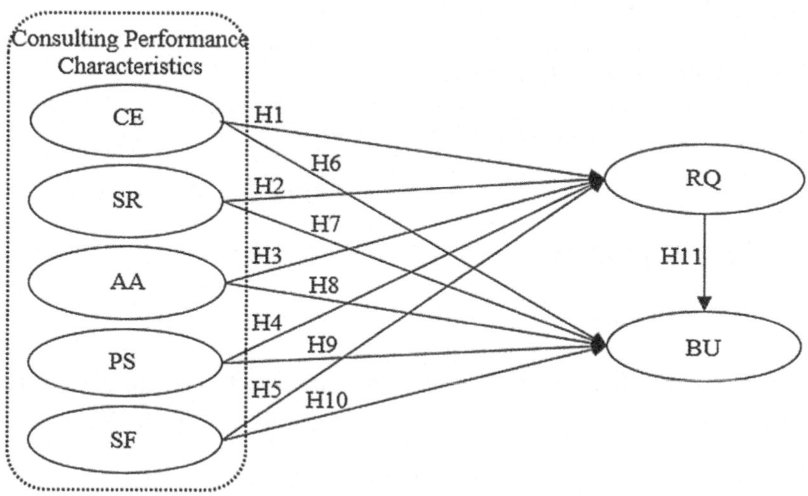

Fig. 1 Research model

performed consulting, and to study what kind of causal relationship between the performance characteristics and the quality of the results was to be investigated.

2.3 Research Hypotheses

In order to search the effect of consulting performance characteristics for the purpose of this research on the quality of consulting results and the utilization of the results, the following hypotheses were established through prior research:

H1 The consultant's expertise will have an affirmative (+) influence on the result quality.
H2 The reliability of consulting services will have an affirmative (+) influence on the result quality.
H3 The acceptance attitude of company toward consulting will have an affirmative (+) influence on the result quality.
H4 The performance suitability of consulting will have an affirmative (+) influence on the result quality.
H5 The fidelity of government support will have an affirmative (+) influence on the result quality.
H6 The consultant's expertise will have an affirmative (+) influence on the business application of the results.
H7 The reliability of consulting services will have an affirmative (+) influence on the business application of the results.
H8 The acceptance attitude of company toward consulting will have an affirmative (+) influence on the business application of the results.
H9 The performance suitability of consulting will have an affirmative (+) influence on the business application of the results.
H10 The fidelity of government support will have an affirmative (+) influence on the business application of the results.
H11 The result quality of consulting will have an affirmative (+) influence on the business application of the results.

2.4 Method of Study

2.4.1 Data Collection

The sample for this study was surveyed on companies with experience in consulting through a consulting program supported by the government. The questionnaire was composed through prior research and was made of a seven-point Likert scale. Among the collected questionnaires, 578 effective cases were extracted and used as research samples.

2.4.2 Operational Definition and Measurement of Variables

Through research on several preceding studies, the measurement variables were composed, and the operational definition of the variables is shown in Table 1.

2.4.3 Statistical Analysis Method

The statistics were analyzed by the structural equation model, and the programs used are SPSS22.0 and AMOS22.0. Firstly, the frequency of demographic characteristics of 578 respondents was analyzed with SPSS. Average, standard deviation, skewness, and kurtosis were extracted through descriptive statistics analysis, and normality was confirmed by satisfying the conditions of skewness 3 or less and kurtosis 10 or less [28]. In addition, variable validity was verified through an exploratory factor analysis method. KMO is 0.864 and Bartlett's sphericity test is 12804.164 ($p < 0.000$). Cumulative total variance explained 76.85%. Cronbach's α value for each factor is within 0.778–0.941, which is reliable. Also, the centralized validity and discriminant validity of the measurement model consisting of seven latent variables were confirmed through confirmatory factor analysis using

Table 1 Operational definition of variables

Variables	Operational definition	Measurement question	Preceding research
Consultant's expertise	The consultant's expertise includes knowledge, analytical skills, strategic thinking skills, documentation skills, and communication skills	5	Bae [1] Ciampi [6] Loukus [7]
Service reliability	It means the expectation that someone else will perform certain actions that are important to me	5	Mayer [8] Choi [9]
Acceptance attitude	When implementing a project, be aware of the project's goals, implement partnerships with the project team, and make the most of consultants	5	Appelbaum [11]
Performance suitability	This indicates the extent to which the performance content is valid or not, and whether the method is appropriate	5	Kim [12]
Support fidelity	Consulting is supported in recognition of the need for external expert help throughout the management of SMEs with insufficient manpower and funds, and funding and project periods appropriate to the size of the enterprise are set	5	Cho [4] Yun [13]
Result quality	Service result quality is the overall impression of the service provider or the relative excellence/inferiority of the service provided	5	Yi [14] Simon [15] Chang [16]
Business utilization	It means how much the consulting service result suggested by the consultant is actually used and applied in the workplace	5	Yi [14] Ryu [17] Yoo [18] Song [19]

AMOS. Finally, the fitness for the structural model was checked and the hypotheses were tested.

3 Results and Discussion

3.1 *Empirical Analysis*

3.1.1 Sample Characteristics

To see the general sample characteristics used in this research, frequency analysis was conducted, and the characteristics of sample can be seen in Table 2. The gender of the respondents was 488 males (84.4%) and 90 females (15.6%). As for the business type, there are 559 corporate businesses (96.7%) and 10 individual businesses (3.3%). And as for the industry, there are 500 manufacturing companies (86.5%) and 78 other types including service industries (13.5%). Lastly, the business period is 247 companies (42.7%) for less than 10 years, 222 companies (38.4%) for less than 20 years, and 109 companies (18.9%) for more than 20 years.

3.1.2 EFA and Reliability Analysis

In order to verify the validity of the variables used in this study, exploratory factor and reliability analyses were performed for seven variables, and the results are shown in Table 3. Principal component analysis (eigenvalue 1.0 or higher) with varimax *was* used. Table 3 shows that the KMO is 0.864 and Bartlett's sphericity test is 12804.164 ($p < 0.001$). The cumulative total variance accounted for 76.85%. Cronbach's α value for each factor is within 0.778–0.941, which is reliable. Previously, there were five measurement items for each variable, but one measurement item for each variable was removed to ensure validity.

Table 2 Sample characteristics

Category		Frequency (person)	Ratio (%)
Gender	Male	488	84.4
	Female	90	15.6
Business type	Corporate business	559	96.7
	Private business	10	3.3
Industry	Manufacturing,	500	86.5
	etc. (service, wholesale and retail, etc.)	78	13.5
Business period	Less than 10 years	247	42.7
	Less than 20 years	222	38.4
	More than 20 years	109	18.9

Table 3 Exploratory factor analysis and reliability analysis

Variables	RQ[1]	SR[2]	PS[3]	SF[4]	CE[5]	AA[6]	BU[7]	Cr's α[8]
E1	0.869							0.941
E2	0.833							
E5	0.809							
E3	0.808							
B5		0.916						0.936
B2		0.906						
B4		0.882						
B1		0.838						
D3			0.849					0.907
D1			0.847					
D2			0.843					
D5			0.623					
F2				0.804				0.919
F5				0.801				
F1				0.748				
F3				0.718				
A3					0.876			0.868
A5					0.847			
A2					0.822			
A1					0.772			
C3						0.885		0.851
C4						0.869		
C5						0.819		
C2						0.731		
H3							0.799	0.778
H4							0.757	
H1							0.731	
H5							0.691	
EV	3492	3432	3277	3033	2925	2.832	2528	
V (%)	12,472	12,257	11,705	10,832	10,445	10,114	9029	
AV (%)	12,472	24,729	36,434	47,265	57,710	67,823	76,852	

KMO = 0.864, Bartlett $x^2 = 12804.164$, $p = 0.000$

Ref.: (1) *RQ* (service) result quality; (2) *SR* service reliability; (3) *PS* performance suitability; (4) *SF* (government) support fidelity; (5) *CE* consultant expertise; (6) *AA* (corporate) acceptance attitude; (7) *BU* business utilization; (8) *Cr's α* Cronbach's α

3.1.3 Confirmatory Factor Analysis

The configuration validity and reliability of the measurement model were verified through confirmatory factor analysis. As shown in Table 4, the standardization coefficient was 0.5 or more, the average variance extracted (AVE) was 0.5 or more, and

Table 4 Goodness of fit for measurement model (concentrated validity)

Path			Non-standardized coefficients	SE	t(CR)	*P*	Standardized coefficient	AVE	CR
E1	<--	CE	1.000				0.916	0.554	0.789
E2	<--		0.865	0.028	30.549	***	0.872		
E3	<--		0.981	0.031	31.506	***	0.885		
B1	<--	SR	1.000				0.824	0.598	0.817
B2	<--		1.123	0.044	25.536	***	0.905		
B4	<--		1.115	0.045	24.779	***	0.875		
F2	<--	AA	1.000				0.868	0.606	0.822
F3	<--		0.998	0.037	27.268	***	0.876		
F5	<--		0.978	0.037	26.665	***	0.863		
C3	<--	PS	1.000				0.907	0.678	0.808
C4_1	<--		0.981	0.101	9.700	***	0.879		
D1	<--	SF	1.000				0.916	0.593	0.814
D2	<--		0.991	0.026	37.897	***	0.932		
D3	<--		0.971	0.027	35.862	***	0.911		
A3	<--	RQ	1.000				0.911	0.693	0.818
A5	<--		0.996	0.072	13,922	***	0.924		
H1	<--	BU	1.000				0.742	0.541	0.702
H3	<--		1.071	0.086	12.469	***	0.789		
CMIN/DF = 3.161(360,395/114) ($p < 0.001$), RMR = 0.072, GFI = 0.936, AGFI = 0.904, NFI = 0.954, IFI = 0.968, TLI = 0.957, CFI = 0.968, RMSEA = 0.061									

the construct reliability (CR) was 0.7 or more, so it was confirmed that it met the criteria.

3.1.4 Discriminant Validity

To check the discriminant validity of the constituent factors, we use a method of verifying by comparing the average variance extracted (AVE) and the correlation coefficient between variables [20], or a method of checking whether the value of the correlation coefficient (ϕ) ± (2 × *SE*) does not include 1 (Woo, Jong-pil 2016). In this paper, the former method was used, and as shown in Table 5, the square root of the average variance extracted (AVE) value of each latent variable was found to be greater than the correlation coefficient with the other latent variables. It can be viewed as appropriate.

Table 5 Comparison of correlation between variables and square root of AVE

Variable	CE	SR	AA	PS	SF	RQ	BU
CE	**0.745**						
SR	0.269	**0.774**					
AA	0.649	0.413	**0.779**				
PS	0.038	−0.004	0.174	**0.824**			
SF	0.643	0.288	0.641	0.075	**0.770**		
RQ	0.068	0.206	0.193	−0.130	0.125	**0.832**	
BU	0.329	0.474	0.514	0.176	0.298	0.328	**0.736**

Table 6 Goodness of fit for structural model

Hypothesis	Path	β	t-value	p	Result
H1	CE→RQ	−0.104	−1.864	0.062	Rejected
H2	SR→RQ	0.167	2.760	0.006	Accepted
H3	AA→RQ	0.253	3.042	0.002	Accepted
H4	PS→RQ	−0.202	−3.660	***	Rejected
H5	SF→RQ	0.026	0.515	0.607	Rejected
H6	CE→BU	0.031	0.982	0.326	Rejected
H7	SR→BU	0.207	5.759	***	Accepted
H8	AA→BU	0.218	4.506	***	Accepted
H9	PS→BU	0.103	3.276	0.001	Accepted
H10	SF→BU	−0.035	−1.254	0.210	Rejected
H11	RQ→BU	0.132	4.871	***	Accepted

3.2 *Results of Hypotheses Test*

3.2.1 Structural Model Analysis

To analyze the fit of the research model, structural model verification was performed. The fit of the structural model is CMIN/DF(Q) = 2.952(342.488/116, ($p < 0.000$)), RMR = 0.082, GFI = 0.939, AGFI = 0.909, NFI = 0.957, IFI = 0.971, TLI =0.961, CFI = 0.971, and RMSEA = 0.058. All indices except RMR, which are affected by the number of samples, show *a* perfect model fit. The route values according to the degree of fitness are shown in Table 6.

3.2.2 Results of Hypotheses Test

The hypotheses testing results are shown in Fig. 2 and Table 7. In Hypothesis 2, reliability of consulting services will have an affirmative (+) influence on the result quality ($p < 0.006$). In Hypothesis 3, the attitude of a company to accept consulting will have an affirmative (+) influence on the result quality ($p < 0.002$). In Hypothesis 7, the reliability of consulting services will have an affirmative (+) influence on the business application of the results ($p < 0.000$). In Hypothesis 8, the attitude of a

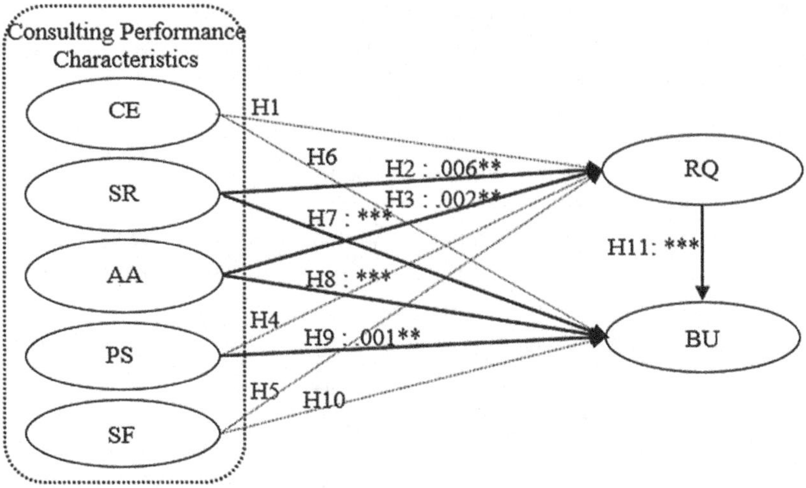

Fig. 2 Results of hypotheses test

Table 7 Results of hypotheses test

H1	The consultant's expertise will have an affirmative (+) influence on the result quality	Rejected
H2	The reliability of consulting services will have an affirmative (+) influence on the result quality	Accepted
H3	The acceptance attitude of company toward consulting will have an affirmative (+) influence on the result quality	Accepted
H4	The performance suitability of consulting will have an affirmative (+) influence on the result quality	Rejected
H5	The fidelity of government support will have an affirmative (+) influence on the result quality	Rejected
H6	The consultant's expertise will have an affirmative (+) influence on the business use of the results	Rejected
H7	The reliability of consulting services will have an affirmative (+) influence on the business application of the results	Accepted
H8	The acceptance attitude of company toward consulting will have an affirmative (+) influence on the business application of the results	Accepted
H9	The performance suitability of consulting will have an affirmative (+) influence on the business application of the results	Accepted
H10	The fidelity of government support will have an affirmative (+) influence on the business application of the results	Rejected
H11	The result quality of consulting will have an affirmative (+) influence on the business application of the results	Accepted

company to accept consulting will have an affirmative (+) influence on the business application of the results ($p < 0.000$). In Hypothesis 9, the adequacy of consulting will have an affirmative (+) influence on the business application of the results ($p < 0.001$). In Hypothesis 11, the result quality of consulting will have an affirmative (+) influence on the business application of the results ($p < 0.000$). The hypotheses of the above study were adopted, and Hypotheses 1, 4, 5, 6, and 10 were rejected, respectively.

4 Conclusion

4.1 Summary of Results and Implications

This research aims to study the influence of consulting performance characteristics on result quality and business utilization in companies that have been consulted using government-supported consulting programs. Through prior research, it is hypothesized that consulting performance characteristics such as expertise, reliability, acceptance attitude, suitability, and fidelity will have an affirmative influence on result quality and work utilization of results, and that result quality will have an affirmative influence on business utilization. As a result of testing the hypotheses, the hypotheses were adopted that the reliability and acceptance attitude of the service will have an affirmative influence on the quality of the result. In addition, the hypothesis that reliability, acceptance attitude, and suitability will have an affirmative influence on business utilization and the hypothesis that result quality will have an affirmative influence on business utilization were adopted. However, the hypotheses that the consultant's expertise, performance suitability, and fidelity of government support will have an affirmative influence on the quality of results and the hypotheses that expertise and fidelity will have an affirmative influence on the use of work were rejected [21–25]. Through previous studies, it was found that the consultant's expertise and the government's faithful support process had an affirmative influence on the quality of results, but these were rejected for this study. Therefore, when only companies using the government-supported consulting program were surveyed, the results show a different aspect from previous studies.

In conclusion, it is understood that the result quality of consulting and the business use of the consulting results differs depending on the purpose, judgment, and attitude of the participating companies, rather than the fidelity of competency of the consultant shown in the government's supporting process. Therefore, I would like to propose two parts for the effective operation of the consulting program supported by the government. First, the consulting support program operating institution needs to focus more on the initial participant selection process. Businesses can perceive government support as simply a system that can be used. In order to improve operational performance, it is necessary to clearly identify the motivation, purpose, and goals of participation so that companies in need of consulting can participate in the initial selection process rather than focusing on the consulting process. Second, it is necessary to seek to strengthen the capabilities of consultants participating in government support programs and improve operational plans. This is because it is common for an expert's competence to influence performance in any field.

4.2 Research Limitations and Future Directions

This study is only for companies that have used government-sponsored consulting projects, and it seems that there will be limitations to extending and applying the conclusions to general companies. Therefore, it is necessary to expand research targeting general companies.

References

1. Bae, Y.S., Ahn, Y.J.: An Empirical Study of the Effects of Consultant Competency on the Performance of Management Consulting: Focusing on the Moderating Effects of the Characteristics of Consulting Firm and Client Firm. Department of Business Administration, Graduate School Dankook University, Republic of Korea, pp.155–170 (2013)
2. Kim, S.H., Leem, W.B.: The impact of consulting service quality on management performance of BSC perspectives. Korean Int'l Account. Rev. **88**, 21–50 (2019)
3. Shapiro, E.C., Eccles, R.G., Soske, T.L.: Consulting: has the solution become part of the problem? MIT Sloan Manag. Rev. **34**(4), 89 (1993)
4. Cho, M.Y.: A Study on the Consulting Performance of SME Consulting Support Project on Acceptance Behavior. Doctoral dissertation. Hansung University, Seoul (2019)
5. Cho, M.Y., You, Y.Y.: A study on the effects of recommendation intention of small and medium-sized enterprise consulting: focused on the comparison between Korean and Chinese small business startups. J. Digit. Contents Soc. **18**(7), 1347–1356 (2017)
6. Ciampi, F.: Exploring knowledge creation pathways in advanced management consulting. Emerging Trends & Issues in Management Consulting: Consulting as a Janus Faced Reality, Inform. Age Pub. pp. 3–40 (2009)
7. Loukus, A.K., Dixon, M.R.: Flawless consulting: a handy tool for behavior analysts and business consultants. In: Block, P. (ed.) Flawless Consulting: a Guide to Getting your Expertise Used. Pfeiffer, San Francisco (2011) (2014)
8. Mayer, R.C., Davis, J.H., Schoorman, F.D.: An integrative model of organizational trust. Acad. Manage. Rev. **20**(3), 709–734 (1995)
9. Choi, J.W.: The Effects of Trustworthiness and Organizational Trust on Organizational Effectiveness. Doctoral dissertation. Graduate School Dongyang University. Gyeongsang-Do (2013)
10. Turner, A.N.: Consulting is more than giving advice. Harv. Bus. Rev. **60**(5), 120–129 (1982)
11. Appelbaum, S.H., Steed, A.J.: The critical success factors in the client-consulting relationship. J. Manag. Dev. **24**, 68–93 (2005)
12. Kim, G.H.: Research on the Accuracy Recognition Quality Control of Computed Tomography Equipment and Improvement of the System. Doctoral dissertation. The Graduate School of Hanseo University. Chungcheong-Do (2018)

13. Yun, S.M.: The Effects of CEO and Organizational Characteristics in Small Manufacturers and Government Support on Decision Making and Business Performance. Ph.D Dissertation, Daejeon University. Daejeon (2019)
14. Yi, Y.J., Lee, J.Y.: A reexamination of the measurement and consequences of service quality: development and application of KS-SQI model. Korean Mark. Assoc. **16**(1), 1–26 (2001)
15. Simon, A., Kumar, V.: Clients' views on strategic capabilities which lead to management consulting success. Manag. Decis. **39**, 362–372 (2001)
16. Chang, L., YongFei, J., Bei, K.: Impact of logistics information service on customer satisfaction: an empirical study with assurance as the moderator (2019)
17. Ryu, J.H.: A Study on the Service Quality of Management Consulting Affecting the Management Performance - Mediating effects of the Consulting Satisfaction and the Consulting Utilization - Ph.D Dissertation, Honam Graduate School University. Jeolla-Do (2017)
18. Yoo, D.W.: The Impact of SME' Leadership and Information Technology on Business Performance. Ph.D Dissertation, Daejeon University. Daejeon (2016)
19. Song, M.H.: A Study on the Learning Satisfaction and Job Application of the e-Learning Safety Education for Educational Personnel. Master's thesis. Korea National Open University, Seoul (2018)
20. Fornell, C., Larcker, D.F.: Structural equation models with unobservable variables and measurement error: algebra and statistics. J. Market. Res. **18**, 382–388 (1981)
21. Mishra, S., Tripathy, H.K., Mallick, P.K., Bhoi, A.K., Barsocchi, P.: EAGA-MLP—an enhanced and adaptive hybrid classification model for diabetes diagnosis. Sensors. **20**(14), 4036 (2020)
22. Mishra, S., Mallick, P.K., Jena, L., Chae, G.S.: Optimization of skewed data using sampling-based preprocessing approach. Front. Public Health. **8**, 274 (2020)
23. Mishra, S., Tripathy, H.K., Mishra, B.K.: Implementation of biologically motivated optimisation approach for tumour categorisation. Int. J. Comput. Aided Eng. Technol. **10**(3), 244–256 (2018)
24. Mishra, S., Mallick, P.K., Tripathy, H.K., Bhoi, A.K., González-Briones, A.: Performance evaluation of a proposed machine learning model for chronic disease datasets using an integrated attribute evaluator and an improved decision tree classifier. Appl. Sci. **10**(22), 8137 (2020)
25. Mallick, P.K., Mishra, S., Chae, G.S.: Digital media news categorization using Bernoulli document model for web content convergence. Pers. Ubiquit. Comput. (2020). https://doi.org/10.1007/s00779-020-01461-9
26. Woodside, A., Lysonski, S.: A general model of traveler destination choice. J. Travel. Res. 27, 8–14 (1989). https://doi.org/10.1177/004728758902700402
27. Bitner, M.J.: Evaluating service encounters; the effects of physical surroundings and employee responses. J. Market. 54, 69–82 (1990)
28. Jaskiewicz, P., Klein, S.: Family Influence and Performance – Theoretical Concepts and Empirical Results. Paper presented at the FERC Conference, Portland, Oregon (2005)

The Impact of Sustainable Management Activities on Corporate Management Performance: Mediating Effect of Competitive Advantage

Mi-Kyoung Jin, Yen-Yoo You, and Sang-Bong Kim

1 Introduction

Competition among companies is intensifying due to the recent economic growth and rapid environmental changes in the market. In addition, as various stakeholders of companies emerge, securing competitiveness is recognized as important in the business environment. In such a business environment, companies are striving to implement sustainable management activities to survive, and currently, sustainable management activities are acting as an important condition for corporate management [1]. Sustainable management activities have already expanded globally since the late 1990s as the importance of environmental issues and the connection with the global economy have increased. Accordingly, companies have converted to sustainable management [2]. Sustainable management comprehensively considers not only the financial performance of a company but also economic, environmental, and social activities that affect management, and pursues nonfinancial performance as well. Today, companies have a responsibility and obligation to properly communicate not only quantitative information but also qualitative information to their stakeholders. Such a sustainable management should be achieved in three directions at the same time, environmentally sustainable scale, socially equal distribution, and economically efficient distribution, not a management strategy with a cross-sectional view centered on economic performance. This means that an integrated and developed concept of management strategy called as TBL (triple bottom line). In this study, from the perspective of TBL, sustainable management activities are considered as economic, environmental, and social variables, and a model is made to study the effects on business performance and the mediating effect of competitive

M.-K. Jin · Y.-Y. You · S.-B. Kim (✉)
Department of Smart Convergence Consulting, Hansung University, Seoul, South Korea
e-mail: threey0818@hansung.ac.kr; brainkim75@hansung.ac.kr

S. R. Samanta et al. (eds.), *Cognitive Computing for Risk Management*,
EAI/Springer Innovations in Communication and Computing,
https://doi.org/10.1007/978-3-030-74517-2_7

advantage. In addition, its purpose is to empirically verify the research through questionnaires.

2 Background

2.1 *Sustainable Management Activities*

The concept of the present sustainable management varies depending on the country or organization and is used interchangeably with ethical management, environmental management, corporate social responsibility, and stakeholder management, and continues to evolve [3]. The triple bottom line (TBL) is a strategic concept for carrying out sustainable management created by John Elkington. This was defined as a concept that reflects economic, environmental, and social interests in an integrated manner in relation to the business activities and stakeholders of a company [4]. Firms cannot exist when economic assets run out. However, in reality, companies face unsustainable conditions much earlier. For this reason, companies must always be able to ensure sufficient funding flows to secure liquidity while still delivering above-average returns to shareholders through economic activities [5]. Environmental management was recognized by early researchers as an act of incurring additional costs of the company, but it is gradually being positively affecting the management performance and is reported to have a close relationship with corporate survival [6]. Corporate social activities should be recognized as an essential factor for a company to survive, as well as economic responsibility, which is the pursuit of profits. It has been studied that the higher the economic performance of social enterprises that practice social activities, the higher the sustainable of social enterprises [7].

2.2 *Competitive Advantage*

What a company can win in the market competition is called competitive advantage, and in order to have a competitive advantage, the company's management resources, that is, technology or management ability, must be differentiated [8]. In addition, competitive advantage means victory in market competition from the perspective of a company. This is obtained when a customer buys a company's product or service on the market. The higher the degree of competitive advantage and the longer a company enjoys it, the higher the management performance can be achieved [9]. Another study defines competitive advantage as a company's value activities, resources, and internal source capabilities. It was argued that valuable activity refers to the act of creating added value, and resources and internal source capabilities are the financial, material, human, and organizational assets of a company [10].

2.3 Management Performance

Management performance as setting strategic goals and performance goals, conducting projects, checking whether the desired outputs and results were achieved, and returning them to decision making [11]. It is suggested that management performance is the sum of all the results achieved in order to realize the goals of each business sector, function, task, group, and individual constituting a company [12]. The method of measuring the management performance of a company has traditionally focused on financial performance through quantified data representing profitability such as sales, net profit, investment return, and market share. However, since financial performance cannot accurately reflect the intangible value of a company, a method of simultaneously measuring not only financial aspects but also nonfinancial performance is used for actual management performance [13].

3 The Research Model and Hypothesis

3.1 Research Model

This study constructed a model to analyze the impact of corporate sustainable management activities on corporate management performance based on the theoretical review of previous studies. Economic, environmental, and social activities were selected as independent variables as corporate sustainable management activities. Sustainable competitive advantage was selected as a parameter, and the company's management performance was selected as a dependent variable. Based on the theoretical background, a hypothesis was established to empirically analyze the relationship between the impact of corporate sustainable management activities on management performance and the mediating effect of sustainable competitive advantage. The research model for this study is shown in Fig. 1.

3.2 Research Hypothesis

H1. A company's sustainable management activities will affect its management performance:

- H1-1. Economic will have a positive (+) effect on management performance.
- H1-2. Environmental activities will have a positive (+) effect on management performance.
- H1-3. Social activities will have a positive (+) effect on management performance.

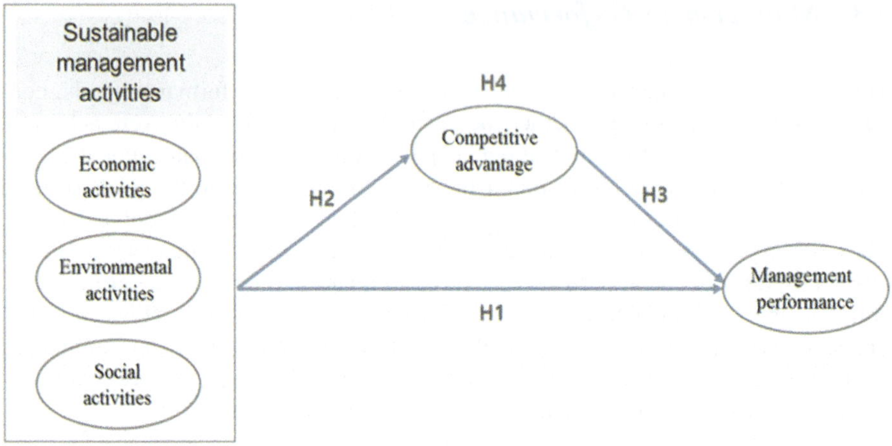

Fig. 1 Research model

H2. A company's sustainable management activities will affect its competitive advantage:

H2-1. Economic activities will have a positive (+) effect on competitive advantage.
H2-2. Environmental activities will have a positive (+) effect on competitive advantage.
H2-3. Social activities will have a positive (+) effect on competitive advantage.

H3. Competitive advantage will have a positive (+) effect on the management performance of a company.
H4. Competitive advantage will play a mediating role between a company's sustainable management activities and management performance:

H4-1. Competitive advantage will play a mediating role between economic activities and management performance.
H4-2. Competitive advantage will play a mediating role between environmental activities and management performance.
H4-3. Competitive advantage will play a mediating role between social activities and management performance.

3.3 *Operational Definition of Variables*

This study measured variables of sustainable management activities, competitive advantage, and management performance according to the research purpose based on previous research results. The operational definition of the variable was summarized in Table 1.

Table 1 Operational definition of variables

Latent variables		Operational definition	Researcher
Sustainable management activities	Economic activities	Businesses should always ensure sufficient funding flow to secure liquidity while delivering steady above-average returns to shareholders through economic activities	[4, 5]
	Environmental activities	Environmental management has a positive effect on management performance. It has a close relationship with the survival of the company	[4, 6]
	Social activities	It has been studied that the higher the economic performance of social enterprises that practice social activities, the higher the sustainable of social enterprises	[4, 7]
Competitive advantage		Competitive advantage means victory in market competition from the perspective of a company	[8–10]
Management performance		Management performance is the result achieved in order to realize the goals of each business, function, task, group, and individual constituting a company. The measurement takes into account both financial and nonfinancial performances	[11–13]

4 Results and Discussion

4.1 Research Model

In this study, 211 samples were collected and the study was conducted, and the frequency analysis of general characteristics of the sample distribution was conducted. As a result of the analysis, 83.4% (176) were male and 16.6% (35) were female. In terms of age, 46–55 years old was the highest with 37.9% (80), followed by 36–45 years old with 34.6% (73). As for the academic background, university graduation was the highest at 59.75% (126).

A descriptive statistics analysis was conducted to confirm the normal distribution of the data [14–18]. As a result of the statistical analysis, the observation variable EN1, which shows extreme values in kurtosis and skewness, was removed. To test the validity of variables, exploratory factor analysis was conducted on sustainable management activities, competitive advantage, and management performance. As a result of factor analysis, the observed variable EN2, which is less than the external loading value of 0.708, was additionally removed.

To test the measurement model of this study, convergent validity, internal consistency reliability, and discriminant validity were tested. Convergent validity is checked based on factor loading values and average variance extraction (AVE), and internal consistency reliability is tested by deriving Cronbach's α and composite reliability.

As a result of the convergent validity and internal consistency reliability test, as shown in Table 2, the factor loading value was at least 0.7, the recommended level, and the AVE was at least 0.5, the recommended level. The internal consistency reliability was tested by deriving Cronbach's α value and composite reliability. Cronbach's α values for latent variables were 0.744–0.897, exceeding the recommended level of 0.7. Composite reliability also showed that all variables were above the recommended level with a value of 0.8 or higher.

To test the discriminant validity, it is confirmed by comparing the correlation between the square root value of the mean-variance extraction (AVE) for each variable and other variable based on Fornell and Larcker's theory. As a result, as shown in Table 3, the largest value among the squared values of the correlation coefficient is 0.674 (competitive advantage ⇔ management performance: r = 0.674). Discriminant validity was obtained because the squared value of the correlation coefficient between all variables did not exceed the AVE value.

Table 2 Results of confirmatory factor analysis

		Convergent validity			Internal consistency reliability		
		Factor loading	Variable reliability	AVE	Cronbach's α	Rho_A (ρ_A)	CR
Latent variables	Observation variables	>0.708	>0.50	>0.50	0.60–0.90	>0.70	0.60–0.90
Competitive advantage	CO1	0.784	0.615	0.709	0.897	0.898	0.924
	CO2	0.882	0.778				
	CO3	0.858	0.736				
	CO4	0.839	0.704				
	CO5	0.845	0.714				
Economic activities	EC1	0.798	0.637	0.624	0.849	0.862	0.892
	EC2	0.707	0.500				
	EC3	0.769	0.591				
	EC4	0.818	0.669				
	EC5	0.850	0.723				
Environmental activities	EN3	0.808	0.653	0.662	0.744	0.771	0.854
	EN4	0.886	0.785				
	EN5	0.741	0.549				
Management performance	MA1	0.762	0.581	0.704	0.894	0.899	0.922
	MA2	0.856	0.733				
	MA3	0.833	0.694				
	MA4	0.894	0.799				
	MA5	0.844	0.712				
Social activities	SO1	0.803	0.645	0.678	0.881	0.885	0.913
	SO2	0.811	0.658				
	SO3	0.876	0.767				
	SO4	0.815	0.664				
	SO5	0.810	0.656				

Table 3 Discriminant validity of the measurement model

Latent variables	Competitive advantage	Economic activities	Environmental activities	Management performance	Social activities
Competitive advantage	**0.842**				
Economic activities	0.631	**0.790**			
Environmental activities	0.645	0.658	**0.814**		
Management performance	0.674	0.609	0.636	**0.839**	
Social activities	0.668	0.476	0.558	0.597	**0.823**

Annotation: the diagonal bold letter is AVE square root

Table 4 Summary of evaluation results of effect size

Latent variables	Competitive advantage	Economic activities	Environmental activities	Management performance	Social activities
Competitive advantage				2.501	
Economic activities	1.818			2.024	
Environmental activities	2.040			2.175	
Management performance					
Social activities	1.497			1.902	

As shown in Table 4, as a result of examining the internal VIF values to evaluate the multicollinearity between each latent variable, all of them were less than 5, indicating that there was no multicollinearity between the latent variables.

Table 5 shows the variance ratio of endogenous potential variables explained by exogenous potential variables, that is, the coefficient of determination (R^2), which means explanatory power. The deciding coefficient for competitive advantage and management performance shows more than the median explanatory power, and the competitive advantage shows a greater explanatory power than management performance.

4.2 *Hypothesis Test Result*

In order to verify the hypothesis of this study, the structural model analysis was analyzed by using the bootstrapping of SmartPLS ver.3.3.2. The following (Table 6) is the result of the estimation of the structural equation model, and all hypotheses have been adopted. The proposed model showed an acceptable model fit.

Table 5 Determination coefficient (R^2) evaluation result

	R^2	Adjusted R^2
Competitive advantage	0.600	0.594
Management performance	0.563	0.554

Table 6 Results of hypothesis test

Hypothesis route	Path coefficient	Standard deviation	*t*-value	*p*-value	Result
Competitive advantage → management performance	0.277	0.091	3.029	0.003	Accept
Economic activities → competitive advantage	0.287	0.068	4.216	0.000	Accept
Economic activities → management performance	0.197	0.064	3.069	0.002	Accept
Environmental activities → competitive advantage	0.232	0.072	3.215	0.001	Accept
Environmental activities → management performance	0.218	0.075	2.922	0.004	Accept

The path model for the impact of corporate sustainable management activities on management performance is shown in Fig. 2.

A specific indirect effect analysis was conducted to verify the mediating effect of competitive advantage in the relationship between corporate sustainable management activities and management performance. As a result of the mediating effect analysis, as shown in Table 7, the mediating effect of competitive advantage was large, and most of the hypotheses were accepted.

5 Conclusion

As a result of the study, sustainability management activities that affect the business performance of a company were analyzed by analyzing previous studies and set as economic, environmental, and social activities. As a result, all factors were found to have a statistically significant effect. Environmental activities (0.218) were found to have the greatest impact on management performance, and economic activities (0.197) and social activities (0.197) were the same. In addition, the impact of sustainable management activities on competitive advantage was in the order of social activities (0.402), economic activities (0.287), and environmental activities (0.232). The competitive advantage set as a parameter was found to have the greatest impact on management performance. In terms of competitive advantage, mediating effects on sustainable activities (economic activities, environmental activities, social activities) were all partially mediated.

The difference between this study and the previous studies is significant in that it studied the mediating effect of a company's competitive advantage between sustainable management activities as an independent variable and management

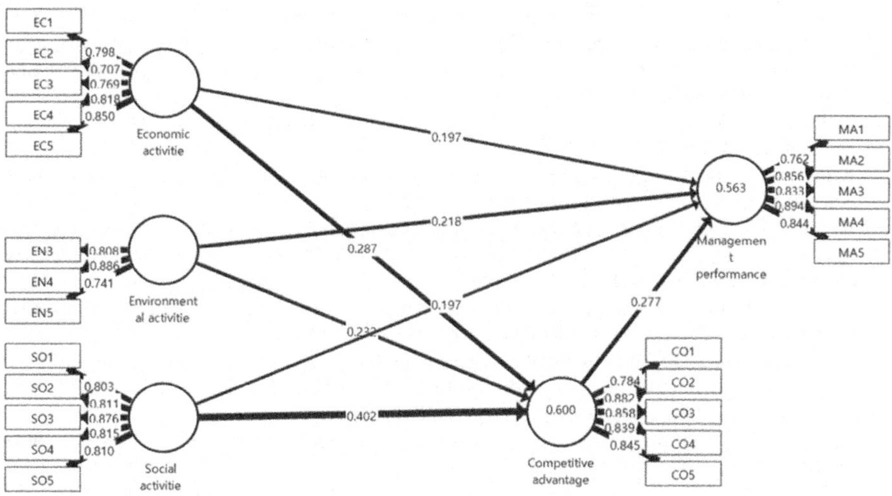

Fig. 2 Path coefficient analysis

Table 7 Result of mediating effect analysis

Hypothesis route	Path coefficient	Standard deviation	*t*-value	*p*-value	Result
Economic activities → competitive advantage → management performance	0.079	0.031	2.530	0.012	Accept
Environmental activities →> competitive advantage → management performance	0.064	0.031	2.098	0.036	Accept
Social activities → competitive advantage → management performance	0.111	0.041	2.737	0.006	Accept

performance as a dependent variable. However, first, the research model and variables were set in a limited range based on previous studies, so the research was conducted in a state where reliability and validity were secured to some extent. Second, there is a limitation in that the research was conducted only on companies located in certain regions. Therefore, in the future research, it is necessary to develop new variables in accordance with recent environmental changes and to conduct research targeting more companies.

References

1. Yoon, J.C., Lee, W.G., Kim, D.H.: A study on the effects of corporate sustainable management activities on innovation in convergence era. J. Digit. Convergence. **13**(4), 115–125 (2015)
2. Kim, G.R., Kim, W.B.: An empirical study on impact of cooperate sustainability management activity on firm's performance: focused on strategic management system. J. Bus. Educ. **29**(5), 155–181 (2015). Available from: http://www.riss.kr/link?id=A101390149
3. Min, J.H., Ha, S., Kim, B.S.: The impact of firms' sustainability management activities on their short-term and long-term values. Korean Manag. Rev. **44**, 713–735 (2015)

4. Kim, K.H., Choi, M.S.: Study for design management of company through analysis of community implementation elements(3C, 3P, 3R) of sustainable development and fusion of triple bottom line. J. Digit. Des. **11**(3), 321–332 (2011). Available from: http://www.riss.kr/link?id=A103877474
5. Dyllick, T., Hockerts, K.: Beyond the business case for corporate sustainability. Bus. Strateg. Environ. **11**(2), 130–141 (2002)
6. Kim, K.: Effects of corporate sustainability management on short and long-term performance. Korean Acad. Soc. Account. **17**(4), 95–118 (2012)
7. An, S.Y., Han, J.S.: A study on the effect of hotels sustainable management efforts on corporate performance. Int. J. Tour. Hosp. Res. **31**(3), 135–149 (2017). Available from: http://www.riss.kr/link?id=A103046160
8. Hall, R.: A framework linking intangible resources and capabilities to sustainable competitive advantage. Strateg. Manag. J. **14**(8), 607–618 (1993)
9. Goo, J.D., Lee, S.H.: The effects of service productivity on competitive advantage and management performance in the hotel industry. Int. J. Tour. Manage. Sci. **21**(3), 113–127 (2006). Available From: http://www.riss.kr/link?id=A5002984
10. Barney, J.B.: Looking inside for competitive advantage. Acad. Manag. Perspect. **9**(4), 49–61 (1995)
11. Park, S.J., Byun, J.Y.: The effects of reciprocity and trust perception on the relationship between corporate sustainable management activities and corporate performance. Korean Manage. Sci. Rev. **33**(2), 49–64 (2016)
12. Kang, I.S.: A study on the structural relationship analysis of management environment, strategy, firm structure and performance in hotel industry. Doctoral dissertation. Kyonggi University. Gyeonggido (2002). Available From: http://www.riss.kr/link?id=T9040068
13. Seo, M.K.: Franchisor's Core Competency and Market Orientation: Theirs Relationship with Environmental Uncertainty, Competitive Advantage, and Business Performance. Doctoral deissertation. Sejong University. Seoul (2012). Available From: http://www.riss.kr/link?id=T12639484
14. Mishra, S., Tripathy, H.K., Mallick, P.K., Bhoi, A.K., Barsocchi, P.: EAGA-MLP—an enhanced and adaptive hybrid classification model for diabetes diagnosis. Sensors. **20**(14), 4036 (2020)
15. Mishra, S., Mallick, P.K., Jena, L., Chae, G.S.: Optimization of skewed data using sampling-based preprocessing approach. Front. Public Health. **8**, 274 (2020)
16. Mishra, S., Tripathy, H.K., Mishra, B.K.: Implementation of biologically motivated optimisation approach for tumour categorisation. Int. J. Comput. Aided Eng. Technol. **10**(3), 244–256 (2018)
17. Mishra, S., Mallick, P.K., Tripathy, H.K., Bhoi, A.K., González-Briones, A.: Performance evaluation of a proposed machine learning model for chronic disease datasets using an integrated attribute evaluator and an improved decision tree classifier. Appl. Sci. **10**(22), 8137 (2020)
18. Mallick, P.K., Mishra, S., Chae, G.S.: Digital media news categorization using Bernoulli document model for web content convergence. Pers. Ubiquit. Comput. (2020). https://doi.org/10.1007/s00779-020-01461-9

A Study on the Impact of Startups of Security Awareness on the Sustainability of Corporate Growth

Eun-Joung Kim and Yen-Yoo You

1 Introduction

When the ICT environment becomes more complex, in the era of the fourth industrial revolution, technology security is becoming increasingly important, but cases of bankruptcy due to technology leaks by SMEs are also increasing. In particular, the survival rate of Korean startups is lower than that of other leading OECD countries. In order to get over the chasm of startups and SMEs based on ICT technology and strengthen their viability, it is necessary to strengthen the entrepreneurship support guide in terms of security.

Information is an important things to determination of performance of a startup company growth [1]. Recognizing that the survival and growth support measures of startups should be further strengthened, the research was conducted to examine the risk impact of startups [2]. In the case of SMEs with weaker infrastructure than large enterprises, technology leakage accidents frequently occur due to deficiency of security awareness, deficiency of technical equipment, and deficiency of professional manpower. The current status of industrial technology leakage and the seriousness of management of Korean small and medium enterprises were examined, and countermeasures were discussed [3]. In order to establish a step-by-step process for SMEs' own industrial security activities to be established and to be realized as security measures to prevent SMEs from leaking industrial technology, the need to present security activities that SMEs can realistically implement was raised [4]. We examined the factors that influence the size of the company and the performance of patent applications on the level of industrial security capabilities of SMEs, and

E.-J. Kim (✉)
Department of Smart Convergence Consulting, Hansung University, Seoul, South Korea

Y.-Y. You
Department of Knowledge Service & Consulting, Hansung University, Seoul, South Korea
e-mail: threey0818@hansung.ac.kr

S. R. Samanta et al. (eds.), *Cognitive Computing for Risk Management*, EAI/Springer Innovations in Communication and Computing, https://doi.org/10.1007/978-3-030-74517-2_8

suggested ways to improve policies to strengthen SMEs' security capabilities [5]. Now, in order for SMEs to grow beyond survival, they must seriously consider information security, which is a side effect of informatization, and at the same time, countermeasures are needed. It was assumed that the perception of information security could have an effect on information security behavior and that such behavior could affect information security performance, moving away from the existing fragmentary view [6]. Among the information security management areas, we intend to come up with a plan to minimize security threats caused by internal organization members that have relatively high uncertainty and are difficult to control. In order to boost the organization's intent to compliance with information security, the relationship between the organization's security efforts and the organization's security interests is presented [7]. Entrepreneurship is also a momentous factor for startups, but it is equally important for the scale-up of a company after startups, and it might be said that the favorability of startups and growth should be expressed in a balanced manner through entrepreneurship [8]. There is no doubt that security training is necessary even if the impact on raising employee security awareness is different. Therefore, by analyzing the requirements for security education from institutional aspects such as laws, the justification for security education was examined, and the contents and methods of education in the security education sector of the technology protection support system were identified [9]. In an omni-channel environment with the advent of the convergence and convergence era, it is very important that information security measures are effectively implemented through voluntary participation of corporate members along with the expansion of technical and administrative information security measures [10]. Applying the theory of fairness and goal setting in the field of information security, we find a mechanism to boost the organization's intent to compliance with information security. In other words, a study was conducted that the setting of security policy objectives has a positive impact on the level of perception of security-related fairness among members of the organization and that fairness has a positive impact on the compliance purpose [11]. As an example, it was found that financial information security compliance affects financial institutions and financial authorities to respond to crises, and such crisis response activities affect trust in financial information security [12]. The aim is to derive policy implications to multiply the effect of startups contributing to economic growth by creating employment through startups and identifying factors that influence the phenomenon of increasing corporate sales. To this end, the basic variables affecting the growth of sales and the rise in the number of workers of the startups were studied for 3 years after the startup [13]. The relationship and role of two factors, strategic orientation to startup performance and entrepreneurial orientation, were referred to in the study [14].

2 Materials and Methods

2.1 Collecting Data

The study currently looked at SMEs' workers with fewer than 100 workers. A questionnaire using a Likert five-point scale measures the impact of workers' security awareness on the sustainability of growth in the startup business.

208 copies of the questionnaire were collected, but 13 inappropriate answers were excluded from the collected questionnaire responses.

2.2 Research Model

Based on prior research, this study analyzes the impact of the organization's formation of information security culture on the continued growth of SMEs. In addition, it is intended to present the organization's behavioral factors to establish the organization's security system, which will be a growth engine, and to prove the relationship. The study models for empirical analysis of this research are shown in Fig. 1

2.3 Research Hypothesis

Based on the research model in Figure 1, the hypothesis has been established for the relations with the formation of an organization's information security culture and the company's sustainable growth. And then, a research hypothesis was established on the mediated effects of a security system in the relationship between the

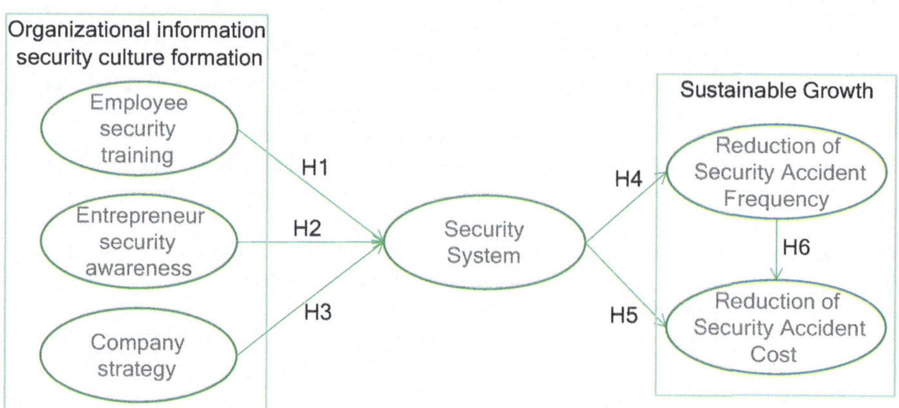

Fig. 1 Research model

organization's information security culture formation and the company's sustainable growth:

H1: Employee security training will have a positive impact on the security system.
H2: Entrepreneur security awareness will have a positive impact on the security system.
H3: Company strategy will have a positive impact on the security system.
H4: Security system will have a positive impact on the decrease of security accident frequency.
H5: Security system will have a positive impact on the decrease of security accident cost.
H6: Decrease of security accident frequency will have a positive impact on the decrease of security accident cost.

2.4 *Operational Definition of Variables*

Based on prior research, this study classified the security culture formation into employee security training, entrepreneur security awareness, and company strategy to construct a questionnaire. To check the impact of this study, the measurement items of the security system and sustainable growth were composed.

The operational definitions of the measurement variables and the composition of configure questionnaires are shown in Table 1.

Table 1 Operational definition of variables

Measurement variables		Operational definitions	Configure questionnaires
Security culture formation	Employee security training	Degree of education and training on information security policies and directions	Five questions
	Entrepreneur security awareness	Level of support and effort from management for information security	Five questions
	Company strategy	Organizational security policy to minimize information system security breaches and accidents	Five questions
Security system	Security system	Organizational security regulations, organizations, procedures, and systems	Five questions
Sustainable growth	Reduction of security accident frequency	Changes in the frequency of security incidents after security system implementation	Five questions
	Reduction of security accident cost	Changes in security incident response costs after security system implementation	Five questions
General question		Sex, age, category of company, number of employees, turnover	Five questions

3 Results and Discussion

3.1 *Statistical Analysis Results*

3.1.1 Technical Statistics Analysis

In the descriptive statistical analysis of the measured variables, the standard deviation was 1.038, which satisfies the condition of 3 or less, and the skewness and kurtosis were 0.174 and 0.346, respectively, and it was determined that the individual measured variables had a normal distribution.

3.1.2 Confirmatory Factor Analysis

For evaluation of measurement models converted into a single-dimensional structure, the reliability and validity of measurement variables were evaluated. The validity was evaluated by dividing into convergent validity, which represents the degree of correlation between measured variables and factors, and discriminant validity, which represents the difference between concepts. Reliability evaluation is considered reliable if Cronbach's α and CR (composite reliability) values of latent variables are 0.7 or higher, and the AVE (average variance extracted) is 0.5 or higher. According to Table 2, it can be evaluated as reliable because all of the measured variables satisfy the standard values.

Focused validity evaluation was performed using the bootstrap method of PLS to analyze the factor loading amount and *t*-value of the measurement items loaded in the composition concept [15–19]. If the factor loading of the measurement items is 0.5 or higher, it is considered that there is an intensive validity. In Table 2, all of the measurement items satisfy the standard value, and the t-value of each factor is 2.628 or more, indicating a significance level of 1%. Since all are significant, it can be evaluated as having concentration validity.

3.1.3 Correlation Analysis

It can be said that discriminant validity exists only that the square root of the mean variance extraction value of each component must be greater than the correlation coefficient value between the concept of composition of the species and the lateral as shown in Table 3.

3.1.4 Multicollinearity Verification

According to analysis, as shown in Table 4, the square root of the mean variance extraction value of each component is greater than the correlation coefficient value of the vertical and horizontal constituent concepts, so discrimination validity exists.

Table 2 Confirmatory factor analysis

Evaluation items	Measurement variables	Estimate	SE	*t*-value	*p*-value	CR	AVE	Cronbach's α
Employee security training (ET)	ET1	0.336				0.595	0.235	0.858
	ET2	0.632	0.48	3.999	***			
	ET3	0.575	0.442	3.898	***			
	ET4	0.697	0.499	4.087	***			
	ET5	0.666	0.502	4.048	***			
Entrepreneur security awareness (EA)	EA1	0.576				0.686	0.309	0.858
	EA2	0.705	0.168	7.064	***			
	EA3	0.757	0.182	7.34	***			
	EA4	0.707	0.174	7.077	***			
	EA5	0.621	0.169	6.519	***			
Company strategy (CS)	CS1	0.676				0.713	0.334	0.860
	CS2	0.644	0.234	5.922	***			
	CS3	0.789	0.179	4.488	***			
	CS4	0.619	0.201	4.402	***			
	CS5	0.476	0.138	1.039	0.299			
Security system (SS)	SS1	0.548				0.500	0.196	0.862
	SS2	0.71	0.126	7.435	***			
	SS3	0.434	0.14	8.476	***			
	SS4	0.422	0.133	7.194	***			
	SS5	0.087	0.119	5.72	***			
Reduction of security accident frequency (AF)	AF1	0.361				0.527	0.202	0.861
	AF2	0.262	0.263	2.628	0.009			
	AF3	0.381	0.329	3.311	***			
	AF4	0.692	0.519	4.108	***			
	AF5	0.66	0.488	4.075	***			
Reduction of security accident cost (AC)	AC1	0.545				0.599	0.2563	0.860
	AC2	0.657	0.19	6.123	***			
	AC3	0.671	0.193	6.184	***			
	AC4	0.709	0.205	6.341	***			
	AC5	0.079	0.12	0.968	0.333			

p-value: $^{*}p < 0.05$, $^{**}p < 0.01$, $^{***}p < 0.001$\

Table 3 Summary of correlation analysis results

	ET	EA	CS
ET	**0.486****		
EA	0.597	**0.557****	
CS	0.476	0.3	**0.578****

Table 4 Multicollinearity verification

		B	Beta	*t*	*p*	Tolerance	VIF
Security culture formation	(Constant)	2.537			0.000		
	ET	0.016	0.020	0.255	0.000	0.807	1.239
	EA	−0.015	−0.018	−0.218	0.000	0.676	1.479
	ST	0.131	0.150	1.663	0.000	0.585	1.709

Next, we would like to examine the possibility of multiple public elections for independent countries. The multicollinearity of independent countries was analyzed by tolerance and variance inflation factor (VIF). In general, when the VIF is 4 or more and the tolerance is 0.1 or less, it is determined that multicollinearity exists.

As a result of the measurement, tolerance of all measurement variables such as employee security training, entrepreneur security awareness, and company strategy was 0.1 or more. The VIF value also showed that all of the measured variables were less than 4, indicating that there was no multicollinearity problem, and it can be evaluated as valid.

3.2 *Hypothesis Verification*

3.2.1 Structural Equation Modeling

The path relationship for the causal model proposed in this study includes reduction of security accident frequency and reduction of security accident cost related to sustainable growth, and consists of employee security training, entrepreneur security awareness, and company strategy related to security culture formation. The causal relationship between routes was evaluated by presenting the route coefficient value and t-value in the analysis results of Fig. 2 and Table 5. After analyzing the structural equation model, the following results were derived.

First, the hypothesis that employee security training has a positive impact on the security system was adopted ($\beta = 0.398$, $p < 0.01$). Second, the hypothesis that entrepreneur security awareness positively affects the security system was adopted ($\beta = 0.297, p < 0.01$). Third, the hypothesis that the company strategy has a positive impact on the security system was adopted ($\beta = 0.247, p < 0.01$). Fourth, the hypothesis that the security system has a positive impact on the reduction of security accident frequency was adopted ($\beta = 0.515$, $p < 0.01$). Fifth, the hypothesis that the security system has a positive impact on the decrease of security accident cost was adopted ($\beta = 0.488$, $p < 0.01$). Sixth, the hypothesis that reduction of security accident frequency positively affects the decline of security accident cost was adopted ($\beta = 0.547, p < 0.01$).

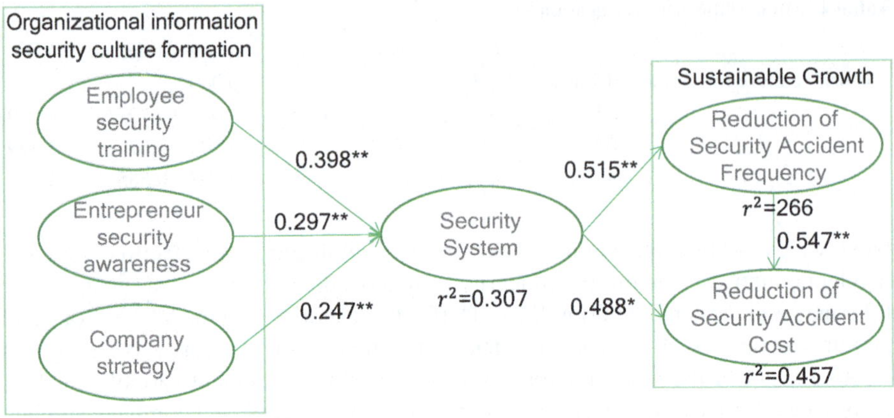

Fig. 2 Structural equation modeling

Table 5 Research model analysis

The hypothesis	Estimate	SE	CR	*p*-value	Results	R2
H1 (ET → SS)	0.398	0.207	2.671	0.008	Accept	0.307
H2 (EA → SS)	0.297	0.079	2.841	0.004	Accept	
H3 (CS → SS)	0.247	0.066	2.497	0.013	Accept	
H4 (SS → AF)	0.515	0.121	3.021	0.003	Accept	0.266
H5 (SS → AC)	0.488	0.153	1.644	0.1	Accept	0.457
H6 (AF → AC)	0.547	0.316	2.988	0.003	Accept	
Model fit $\chi2 = 921.931(df = 399, p = 0.000)$, $\chi2/DF(Q\ \chi) = 2.311$, GFI = 0.767, AGFI = 0.728, NFI = 0.558, RMSEA = 0.082						

p-value: $*p < 0.05$, $**p < 0.01$, $***p < 0.001$

3.2.2 Mediating Effect Analysis

The security system has been found to mediate the impact of employee security training on the reduction of security accident frequency and reduction of security accident cost as shown in Table 6.

It was found that the security system had a greater impact on the reduction of security accident frequency than the reduction of security accident cost.

4 Conclusion

The results obtained through this study classified employee security training, entrepreneur security awareness, and company strategy as independent variables based on previous studies based on the cultivation of security awareness of startup companies. All three variables showed a positive (+) impact on the security system, and the security system showed a statistically significant positive (+) impact on sustainable

Table 6 Mediating effect analysis (security system)

Model	Indirect effects	p-value	Results
ET->SS->AF	0.205	0.002	Accept
EA->SS->AF	0.153	0.032	Reject
CS->SS->AF	0.127	0.062	Reject
ET->SS->AC	0.194	0.002	Accept
EA->SS->AC	0.145	0.032	Reject
CS->SS->AC	0.120	0.062	Reject

growth. Thus, we analyzed empirically that all hypotheses were accepted. In addition, the security system has been found to mediate the impact of employee security training on the reduction of security accident frequency and reduction of security accident cost.

Although it was conducted on the entire startup, it is expected to be useful for startups in the ICT sector because it is a real ICT technology-based startup. However, it is hard to see that the consequence of this research represents the characteristics of all ICT-based startups.

Therefore, more research is needed by dividing the scope into different types of SME industry, industry, etc., so it is necessary to do more research by classifying the scope.

References

1. Lee, C.W., Seoh, C.S.: A exploratory study on the importance of managerial factors of Korean Ventures through organizational stage. Asia Pac. J. Small Bus. **28**(2), 3–29 (2006)
2. Kwak, D.C., Joo, Y.H., Cho, B.H.: A study on influential factors of survival rates: focused on youth start-ups. Asia Pac. J. Small Bus. **38**(4), 77–94 (2016)
3. Gong, B.W.: The situation and security measures of industrial technology security management of SMEs. J. Korean Soc. Priv. Secur. **18**(1), 1–26 (2019)
4. Lee, M.H.: A study on the developmental activation plan of local medium and small firms -focussed on industrial security. Korean Local Auton. Res. **15**(2), 141–159 (2013)
5. Noh, M.S., Lee, S.Y.: Explaining industrial security of SMEs in Korea: an ordered logit analysis. Korean Public Adm. **44**(3), 1–10 (2010)
6. Baek, M.J., Sohn, S.H.: A study on the effect of information security awareness and behavior on the information security performance in small and medium sized organization. Small Bus. Res. **33**(2), 113–132 (2011)
7. Hwang, I.H., Kim, D.J., Kim, T.H., Kim, J.S.: Effect of security culture on security compliance and knowledge of employees. Inf. Syst. Rev. **18**(1), 1–18 (2016)
8. Kim, K.W., Kim, M.S.: An empirical analysis of influence of corporate entrepreneurship on business performance from the viewpoint of SMEs' growth. Asia Pac. J. Bus. Ventur. Entrep. **12**(5), 13–28 (2018)
9. S.L. Park: In terms of technology protection, the frequency of security training Impact on the cultivation of employee security awareness. Master's thesis. Dankook University. Seoul (2019)
10. Han, J.Y., Kim, Y.J.: Investigating of psychological factors affecting information security compliance intention: convergent approach to information security and organizational citizenship behavior. J. Digit. Convergence. **13**(8), 133–144 (2015)

11. Hwang, I.H., Kim, S.W.: A study on the influence of organizational information security goal setting and justice on security policy compliance intention. J. Digit. Convergence. **16**(2), 117–126 (2018)
12. Yoon, I.H., Kwon, S.D.: A study on the effect of information security compliance and crisis management on information security trust. Inf. Syst. Rev. **17**(1), 141–169 (2015)
13. Park, J.B., Kim, J.H., Yang, H.B.: Factors affecting the growth of corporate start-ups focused on sales performance and job creation. Asia Pac. J. Small Bus. **32**(3), 109–128 (2010)
14. Shim, Y.S., Seo, J.H.: A study on entrepreneurial orientation and startup performance: mediating effect of strategic orientation. Int. J. Emerg. Multidiscip. Res. **3**(4), 33–41 (2019). https://doi.org/10.22662/IJEMR. 3.4.033
15. Mishra, S., Tripathy, H.K., Mallick, P.K., Bhoi, A.K., Barsocchi, P.: EAGA-MLP—an enhanced and adaptive hybrid classification model for diabetes diagnosis. Sensors. **20**(14), 4036 (2020)
16. Mishra, S., Mallick, P.K., Jena, L., Chae, G.S.: Optimization of skewed data using sampling-based preprocessing approach. Front. Public Health. **8**, 274 (2020)
17. Mishra, S., Tripathy, H.K., Mishra, B.K.: Implementation of biologically motivated optimisation approach for tumour categorisation. Int. J. Comput. Aided Eng. Technol. **10**(3), 244–256 (2018)
18. Mishra, S., Mallick, P.K., Tripathy, H.K., Bhoi, A.K., González-Briones, A.: Performance evaluation of a proposed machine learning model for chronic disease datasets using an integrated attribute evaluator and an improved decision tree classifier. Appl. Sci. **10**(22), 8137 (2020)
19. Mallick, P.K., Mishra, S., Chae, G.S.: Digital media news categorization using Bernoulli document model for web content convergence. Pers. Ubiquit. Comput. (2020). https://doi.org/10.1007/s00779-020-01461-9

A Case Study on the Association of Supporting Large Enterprises for Win-Win Growth on the Internal Effect of Employees of Primary Partner SMEs in Korea

Ji-Yeon Go, Joo-Young Lee, Ha-Eun Jang, and Yen-Yoo You

1 Introduction

The South Korean government has proposed a win-win growth to deal with polarization issue between large enterprises and SMEs (small and medium enterprises) due to rapid economic growth and to grow for the economic development of large enterprises and SMEs [1]. Win-win growth is an economic growth policy that focuses on growth and distribution at the same time and grows together through cooperation through fair trade [2]. On 2010, the Korea government announced win-win growth plan of large enterprises and SMEs and launched the committee for win-win growth. The committee (Korea Commission for Corporate Partnership) has estimated and published Win-Win Index to promote win-win growth at private sectors [1]. For win-win growth with SMEs, large enterprises have established and implemented fair trade relations with SMEs. And large enterprises supported the development of SMEs through financial support, education, training, technical support, domestic and overseas market support, and support for the improvement of working environment of SMEs' employees [3]. Win-win growth activities by large enterprises and SMEs had a positive impact on the profitability of SMEs, leading to improved technology development and management of SMEs [4]. In addition, supporting SMEs of large enterprises for win-win growth became a driving force to enable sustainable growth of enterprises by enhancing the overall financial

J.-Y. Go
Knowledge Service & Consulting, Hansung University, Seoul, South Korea

J.-Y. Lee · H.-E. Jang
Healthcare Management, Eulji University, Seoul, South Korea

Y.-Y. You (✉)
Division of Smart Management Engineering/Smart Convergence Consulting, Hansung University, Seoul, South Korea
e-mail: threey0818@hansung.ac.kr

S. R. Samanta et al. (eds.), *Cognitive Computing for Risk Management*, EAI/Springer Innovations in Communication and Computing,
https://doi.org/10.1007/978-3-030-74517-2_9

performance and image of large enterprises [5]. This shows that win-win growth activities have a positive impact on both large enterprises and SMEs.

The study found out how supporting SMEs of large enterprises for SMEs has an impact on the internal effects of employees of primary partner SMEs who are doing business with large enterprises that supporting SMEs for win-win growth of large enterprises and SMEs. "Job Satisfaction," "Organizational Identification," and "Organizational Trust" are utilized as variables to measure the internal effects of employees [6]. "Job Satisfaction" is described as the positive emotional state resulting from work experience evaluation or employees' work [7]. According to prior research, "Job Satisfaction" is a critical factor in retaining and attracting skilled employees, and it is an essential aspect for firms to gain a competitive advantage in all sectors, given the central role that employees play in business success [8, 9]. "Organizational Identification" is defined as an awareness of oneness with or belongingness to an organization/group in that the individual notes herself or himself as a symbol of the organization and thinks that the organization's failures and successes are her/his own [10]. Employees that have a stronger sense of "Organizational Identification" tend to have lower turnover intention for job than employees that do not strongly have "Organizational Identification" [11]. "Organizational Trust" is expressed as a general or individual collective position that groups or individuals strive in good faith to act on promises and benefit from the promises made in the relationships they experience [12]. It affects cooperation easier in organizations formed of individuals that come together for the same purpose or to achieve a duty and has a positive effect on organizational relations through communication [13]. In addition, "Organizational Trust" is important for doing work and management functions that bring people together for a one purpose [13].

Most previous study on win-win growth focuses only on external aspects such as financial performance of enterprises [14]. However, internal effect studies on how win-win growth is affecting employees of enterprise who drive the enterprise are very insufficient. This case study is to identify the degree of increase in the internal effects ("Job Satisfaction", "Organizational Identification", "Organizational Trust") of SMEs' employees of primary business partners of Large Enterprises through supporting SMEs of Large Enterprises for Win-Win Growth, and to determine whether differences exist depending on Sex, Age, Rank, Position, Enterprise size and Area.

2 Materials and Methods

This study was conducted based on basic data obtained through a web survey of employees of primary partner of large enterprises. The sampling method was used to judgmental sampling method for the primary partner (SMEs) of large enterprises that were deemed to be most suitable for the purpose of the survey. A total of 250 copies of the questionnaire were collected, and 202 copies were used as final data, excluding insincere responses. The research period was conducted for a total of 20 days from November 1 to November 19, 2019.

2.1 Variable

Sociodemographic characteristics (enterprise area, sex, age, rank in enterprise, position in enterprise, enterprise size, industry) were included in the study model. Enterprise area was classified as "suburban" and "urban." Sex was classified as "male" and "female." Age was categorized as "30 under," "30 or over to 40 under," "40 or over to 50 under," and "50 or over." Rank in enterprise was classified as "staff," "executive," and "chief." Position in enterprise was categorized as "management support," "product/technique," "R&D," and "sales/marketing." Enterprise size was classified as "small," and "medium." Industry was categorized as "service," "manufacturing," and "etc."

The dependent variable, support for small and medium enterprise in Korea's large enterprises, was assessed using six questions, measured on the five-point Likert scale of "very disagree," "disagree," "normal," "agree," and "very agree." The list of a question of the survey was based on the evaluation of the win-win growth index by the commission for win-win growth; the "supporting SMEs of large enterprises" was measured with a total of six questions using the questionnaire composing of Lee [15].

The "Job Satisfaction," "Organizational Identification," "Organizational Trust" dependent variable was used in the study model. "Job Satisfaction" was assessed using five questions, measured on the five-point Likert scales of "very disagree," "disagree," "normal," "agree," and "very agree." The list of the question of the survey was based on the Minnesota Satisfaction Questionnaire (1967); the "Job Satisfaction" was measured with a total of five questions using the questionnaire composing of Kim, S. Y. (2015), Park, E. K. (2018), and Yiyi Hui (2019) [16–19]. "Organizational Identification" was assessed using four questions, measured on the five-point Likert scale of "very disagree," "disagree," "normal," "agree," and "very agree." The list of the question of the survey was based on the "Mael and Ashforth" (1995); the "Organizational Identification" was measured with a total of four questions using the questionnaire composing of Cho, D. H. (2015) and Park, E. K. (2018) [18, 20, 21]. "Organizational Trust" was assessed using five questions, measured on the five-point Likert scale of "very disagree," "disagree," "normal," "agree," and "very agree." The "Organizational Trust" was measured with a total of five questions using the questionnaire of Zi, S. K. (2006), Woo, J. S. (2016), and Lee, J. K. (2019) [22–24].

2.2 *Statistical Analysis*

Descriptive analysis was conducted to confirm the correlation between the internal effects of employees of primary cooperative SMEs to supporting SMEs' large enterprises. To analyze the internal effect in participant characteristics, the correlation test was carried out, and the correlation of positive (+) was all confirmed. The results of Cronbach's alpha values, which measure the reliability for "Job Satisfaction," "Organizational Identification," and "Organizational Trust," were all located between 0.7 and 0.9, thus securing desirable reliability of 0.7 or higher. The multi-linear regression analysis was added to identify associated with supporting SMEs of Large Enterprises by internal effects. The normal distribution of the dependent variables, the assumption of linear regression, was confirmed through the ANOVA test. The collected data were analyzed using the statistical program IBM SPSS software (Version 23.0, Chicago, IL, USA). The threshold for statistical significance was 0.05 (two-tailed).

3 Results and Discussion

Sociodemographic characteristics of the participants are shown in Table 1. A total of 202 primary partner (SMEs) employees of large enterprises participated the survey. Of these, 128 (63%) were male and 74 (37%) were female. The age are 24

Table 1 Participant's sociodemographic characteristics

Variable	Category	*N*	%
Sex	Male	128	63%
	Female	74	37%
Age	Under 30	24	12%
	30–39	42	21%
	40–49	82	41%
	Over 50	54	27%
Rank	Staff	54	27%
	Executive	126	62%
	Chief	22	11%
Position	Management support	76	38%
	Product/technique	18	9%
	R&D	8	4%
	Sales/marketing	100	50%
Enterprise size	Small	42	21%
	Medium	160	79%
Area	Suburban	20	10%
	Urban	182	90%
Total		202	100%

people (12%) under 30 years old, 42 people (21%) 30–39 years old, 82 people (41%) 40–49 years old, and 54 people (27%) over 50 years old. The rank are 54 people (27%) in staff, 126 people (62%) in executive, and 22 people (11%) in chief. The position are 76 people (38%) in management support, 18 people (9%) in product/technique, 8 people (4%) in R&D, and 100 people (50%) in sales/marketing. The enterprise size are 42 people (21%) in small enterprise and 160 people (79%) in medium enterprise. The area are 20 people (10%) in suburban and 182 people (90%) in urban.

Supporting SMEs of large enterprises of mean and standard deviation are shown in Table 2. The mean of supporting SMEs of large enterprises is 3.6; standard deviation is 0.05 (Table 3).

The correlation between support for SMEs of Large Enterprises and "Job Satisfaction" (.729) of SMES, "Organizational Identification" (.756), and "Organizational Trust" (.678) were analyzed through 0.05 (two-tailed) and positive correlation was confirmed. It is shown in the table 3. Based on this, it was confirmed that none of the internal effects could be overlooked because they were related to each other.

Results of the linear regression analysis by supporting SMEs of large enterprises are shown in Table 4. "Job Satisfaction" (ß, 0.206; 95%CI, 0.041–0.327; $P = 0.012$), "Organizational Identification" (ß, 0.267; 95%CI, 0.114–0.408; $P = 0.001$), and "Organizational Trust" (ß, 0.389; 95%CI, 0.190–0.445; $P = 0.000$) were correlated with supporting SMEs of large enterprises. This showed that supporting SMEs of large enterprises and internal effects ("Job Satisfaction," "Organizational Identification," "Organizational Trust") positively correlated. Among sex, females (ß, 0.160; 95%CI, 0.093–0.374; $P = 0.001$) were positively correlated with supporting SMEs of large enterprises. Among position, product/technique (ß, −0.097; 95%CI, −0.479–0.000; $P = 0.050$) and sales/marketing (ß, −0.140; 95%CI, −0.479–0.000; $P = 0.004$) were negatively correlated with supporting SMEs of large enterprises.

In the case of females, they were more positive than males about supporting primary business partners of large corporations. It is believed that the primary partner support of large enterprises includes improvements in educational resource working conditions, which are difficult for ordinary SMEs to realize on their own, so they feel the practical benefits compared to men [25]. In addition, according to the Korea National Statistical Office in 2019, 21% of men and 79% of women employed in Korean companies took child-care leave [26]. The results are highly related to Korea's vertical corporate culture. The large companies are oriented toward a horizontal culture and a performance-oriented system, SMEs still form a vertical culture [27], especially for men, it is difficult to use it even if benefits such as child-care leave because scores such as job evaluations and job evaluations are

Table 2 Mean and standard deviation of independent variable

Variable	Mean	SE
Supporting SMEs of large enterprises	3.6	0.05

Table 3 Correlation among internal effects

			Support of large enterprise	Job Satisfaction	Organizational Trust	Organizational Identification
Supporting SMEs of large enterprises		Pearson correlation	1	0.729**	0.756**	0.678**
		Sig. (two-tailed)		0.000	0.000	0.000
		N	202	202	202	202
Internal effects	Job Satisfaction	Pearson correlation	0.729**	1	0.808**	0.758**
		Sig. (two-tailed)	0.000		0.000	0.000
		N	202	202	202	202
	Organizational Identification	Pearson correlation	0.756**	0.808**	1	0.727**
		Sig. (two-tailed)	0.000	0.000		0.000
		N	202	202	202	202
	Organizational Trust	Pearson correlation	0.678**	0.758**	0.727**	1
		Sig. (two-tailed)	0.000	0.000	0.000	
		N	202	202	202	202

related to promotions. Therefore, support for primary suppliers of large companies is important, but support for changing the fundamental culture is needed, and it is necessary to calculate the score of evaluation of win-win growth of large companies.

The positions case, "Chief" of primary partners at large companies did not significant to support from large companies, but the fact that all CI values are negative (−) suggests a lot. This can be judged that "Chief" feel very insufficient for support from large corporations compared to employees. "Chief," who actually run the company, feel the lack of support from large companies which means that they believe that the support of large companies themselves is a formality for the scores of win-win cooperation between large companies. In other words, it is perceived as a troublesome task added to the needs of large companies without providing any practical help to SMEs. In this context, "product/technique" and "sales/marketing" departments were feeling a lack of support from large enterprises compared to "management support." The contents of win-win cooperation include items such as technical support, technology protection, financial support, and business sales support [25], but indicate that the level of support from large companies is lower than expected by employees of SMEs. Since the majority of SMEs have "product/technique," called

Table 4 Results of the linear regression analysis by supporting SMEs of large enterprises ($N = 202$)

					95% CI		
Variable	Category	ß	SE	*P*-value	LL	UL	VIF
Internal effects	Job Satisfaction	0.206	0.073	0.012	0.041	0.327	3.781
	Organizational Identification	0.267	0.075	0.001	0.114	0.408	3.348
	Organizational Trust	0.389	0.065	0.000	0.190	0.445	3.609
Sex	Male	ref					
	Female	0.160	0.071	0.001	0.093	0.374	1.376
Age	Under 30	ref					
	30–39	0.104	0.132	0.173	−0.080	0.441	3.353
	40–49	0.105	0.159	0.347	−0.164	0.464	7.097
	Over 50	0.098	0.166	0.352	−0.173	0.483	6.319
Rank	Staff	ref					
	Executive	−0.015	0.118	0.849	−0.254	0.210	3.782
	Chief	−0.113	0.156	0.103	−0.564	0.052	2.761
Position	Management support	ref					
	Product/technique	−0.097	0.121	0.050	−0.479	0.000	1.395
	R&D	0.015	0.163	0.738	−0.267	0.375	1.172
	Sales/marketing	−0.140	0.068	0.004	−0.332	−0.063	1.355
Enterprise size	Small	ref					
	Medium	0.021	0.088	0.683	−0.138	0.211	1.501
Area	Urban	ref					
	Suburban	−0.036	0.108	0.430	−0.298	0.127	1.208

root industries, and the "sales/marketing" department is a core department that is intuitively related to the sales of companies, the lack of support from large companies by the employees of the department shows the current level of win-win growth and its limitations.

The primary partner support of large companies shows a positive correlation between internal effects such as "Job Satisfaction" of SMEs, the "Organizational Identification," and "Organizational Trust." Also "Job Satisfaction," "Organizational Identification," and "Organizational Trust" all had high-volume correlation with supporting variables from large corporations. And coefficient of determination (R2) of supports from large enterprises on SMEs was in order by "Organizational Trust" (R2 = 0.627), "Organizational Identification" (R2 = 0.609), "Job Satisfaction" (R2 = 0.569). Therefore, it can be confirmed that the support of SMEs for win-win growth is positively improving the internal effectiveness of SMEs' employee. In light of the "Herzberg, F theory," the results of the amount of "Job Satisfaction" indicate that large companies' support for SMEs actually serves as a motivator for individuals, not as a temporary improvement in the surrounding environment (motivator factor) [28]. The positive (+) result of the "Organizational Identification" means that employees of SMEs have become more attached to the organization through the support of SMEs by large companies [10]. This is because the title of a partner plays a role in enhancing the reputation of the SMEs in question, According to the prior study, in general, the more members perceive their organization as having a better external reputation [29]. A positive (+) result in "Organizational Trust" means that SMEs' employees are judged to have been treated more fairly by their organizations than before [12]. This is because the title of a partner plays a role in enhancing the reputation of the SMEs in question. According to the prior study, in general, "Organizational Identification" becomes much solid what members perceive their organization as having a better external reputation [29]. The positive (+) result of "Organizational Trust" indicates that SMEs' employees believe that the support of SMEs by large enterprises has made the treatment they receive from the organization fairer than before [12].

Correlation analysis shows that support for SMEs by large enterprises has shown a high (+) relationship between "Job Satisfaction," "Organizational Identification," and "Organizational Trust." This means that there is a positive interaction between the support of SMEs and the internal effects of large companies that are currently being implemented as part of win-win growth. The increase in internal effects can lead to the satisfaction of workers at SMEs, which lead to the extension of their service life, which can greatly help solve the labor shortage, the biggest problem for SMEs, and narrow the gap between wages and welfare with large companies, so the importance of this is very high. What should be noted here is that there is also a high correlation between internal effects. The increase in "Job Satisfaction" leads to "Organizational Civic Behavior" [30], and "Organizational Identification" increases the sense of unity in the company, thereby better accommodating the values of the organization and contributing to the achievement of the organization's goals through job immersion [10], and increasing the awareness of fairness in "Organizational Trust," so that employees' task attitudes and performance are highly correlated [13]. Therefore, it means that the support of large enterprises through win-win growth is not carried out with the aim of improving the internal effects of a single category, but it is necessary to support all areas of SMEs in an integrated manner, and policy makers should consider these areas and draw up policies.

This study has some limitations. First, this study is based on a cross-sectional design, which limits association between "support of large enterprise" and "internal effect of SMEs employees." Second, self-reported data on internal effect of SMEs may not provide exact information because of recall and social desirability biases. Third, SMEs not supported by large enterprise were not considered in this analysis, and since it is a special policy only for Korea, it is difficult to apply it collectively to other countries.

Despite these limitations, this study has several strengths. This study has focused on SMEs of employee in Korea. In the case of a prior study, a study was conducted on external effect (sales, net income, etc.), but there was no research of internal effect. Several internal effect variables were used to determine the correlation between each variable and the support of large enterprises [31–35]. This study can be used as data on performance evaluation of win-win growth programs and can remind policy makers of the importance of increasing internal effects, not external ones.

4 Conclusion

This study has showed that relation of internal effect ("Job Satisfaction", "Organizational Identification", "Organizational Trust") and support of large enterprise by SMEs employee. This chapter found that females were more aware of support from large companies than male, but CI was all negative for "chief" compared to "staff", and "Product/Technique" and "Sales/Marketing" departments were less aware of support from large companies than "Management Support" were less aware of support from large companies. On the other hand, it was confirmed that the primary business partner support of large companies is positively improving internal effects and that there is a high correlation between internal effects. Therefore, the policy makers have confirmed the need to compensate for the lack of support for win-win growth, and should play a role in changing the corporate culture of SMEs, not in the form of formal work, and should present policies that can enhance the overall effectiveness of the company in order to enhance its internal effectiveness.

References

1. Park, H.-H., Kang, D.-N., Lee, S.-Y.: A study on a mutual win-win growth strategy for Korean supermarkets. J. Distrib. Sci. **12**(3), 43–53 (2014)
2. Kim, K.-T., Lee, J.S., Kang, N.J., Management: Empirical study on inter-firm diffusion and firms' performance for win-win growth culture in supply chain. J. Inf. Technol. Appl. Manag. **22**(1), 153–163 (2015)

3. Baek, J., Kwon, S.B., Choi, B.J.K.M.R.: Evaluating the impact of win-win growth policy announcements between large firms and SMEs on the market value of firms. Knowl. Manag. Soc. Korea. **13**(5), 139–160 (2012)
4. Song, H.J., Lee, M.Y.: The effects of large firms` win-win growth activity on smes` business performance: a case study on the global hidden champion nurturing program by Samsung electronics. Account. Res. **25**(5), 113–139 (2016)
5. Min, J.H., Kim, B.S.: The impact of win-win growth effort of large firms on their financial performance. Korean Manag. Sci. Rev. **30**(2), 79–95 (2013)
6. Choi, C.S.: A Study on the Internal Effects of Corporate Social Responsibility. Doctoral dissertation. Hanyang University. Seoul (2011)
7. Zhang, X., Ma, L., Xu, B., Xu, F.J.I., Management: How social media usage affects employees' job satisfaction and turnover intention: an empirical study in China. Inf. Manag. **56**(6), 103136 (2019)
8. Stamolampros, P., Korfiatis, N., Chalvatzis, K., Buhalis, D.J.T.M.: Job satisfaction and employee turnover determinants in high contact services: insights from Employees' Online reviews. Tour. Manag. **75**, 130–147 (2019)
9. Navajas-Romero, V., Díaz-Carrión, R., Casas-Rosal, J.C.J.I.J.o.H.M.: Comparing working conditions and job satisfaction in hospitality workers across Europe. Int. J. Hosp. Manag. **90**, 102631 (2020)
10. O'Neill, J.L., Gaither, C.A.J.R.i.S., Pharmacy A: Investigating the relationship between the practice of pharmaceutical care, construed external image, organizational identification, and job turnover intention of community pharmacists. Res. Social Adm. Pharm. **3**(4), 438–463 (2007)
11. Harris, G.E., Cameron, J.E.: Multiple dimensions of organizational identification and commitment as predictors of turnover intentions and psychological Well-being. Can. J. Behav. Sci./ Revue canadienne des sciences du comportement. **37**(3), 159–169 (2005)
12. Basit, G., Duygulu, S.J.C.: Nurses' organizational trust and intention to continue working at hospitals in Turkey. Collegian. **25**(2), 163–169 (2018)
13. Özgür, G., Tektaş, P.J.A.N.R.: An examination of the correlation between nurses' organizational trust and burnout levels. Appl. Nurs. Res. **43**, 93–97 (2018)
14. Shin, Y.-M., Lee, C.-H.J.J.o.D.C.: An empirical study on the effect of win-win growth activity on management performance. J. Digit. Convergence. **16**(7), 71–78 (2018)
15. Lee, H.S.: A study on the determinants and outcome variables on CSR of small and medium sized enterprises(SMEs). Doctoral dissertation. Dankook University. Gyeonggi-do (2018)
16. Weiss, D.J., Dawis, R.V., England G.W.J.M.s.i.v.r.: Manual for the Minnesota satisfaction questionnaire (1967)
17. Kim, S-Y.: A Study on the Effects of Internal Reputation of Organization on the Job Behavior: Focused on the Child Care Facilities in Seoul. Doctoral dissertation. Hansung University. Seoul (2015)
18. Park, E-K.: The Effect of Casino Employees' CSR Participation on Job Satisfaction : Focusing on Mediating Effect of Organizational Identification. Master's thesis. Kyung Hee University. Seoul (2018)
19. Yiyi Hui.: CSR perception and organizational identification: the moderating effect of CSR attribution. Master's thesis. Kyung Hee University. Seoul (2019)
20. Mael, F.A., Ashforth, B.E.J.P.p.: Loyal from day one: biodata, organizational identification, and turnover among newcomers. Pers. Psychol. **48**(2), 309–333 (1995)
21. Cho, D-H.: A cross-level analysis about Relationships among Organizational identity, Organizational identification, and Member's Behavior. Doctoral dissertation. Chosun University. Gwangju (2015)
22. Ji, S.-G.: The effect of social responsibility activities of service companies on the organizational trust and service infiltration of employees. J. Korean Manag. Assoc. **19**(5), 1867–1893 (2006)
23. Woo, J.S.: Effects of Corporate Social Responsibility on Organizational Trust and Organizational Commitment in Airlines. Master's thesis. Kyonggi University. Seoul (2016)

24. Lee, C. K.: The impacts of corporate social responsibility activities on employees' organizational trust and organizational commitment. Master's thesis. Kangwon National University. Gangwon-do (2019)
25. Fair Trade Commission.: Data for the implementation evaluation of the Fair Trade Agreement and the revision of the standards. Seoul: Fair Trade Commission (2020). Available From: http://gowith.ftc.go.kr/enview/board/fileMngr?cmd=down&boardId=notice&bltnNo=11579503350689&fileSeq=1&subId=sub06
26. Statistics Korea.: Status of maternity and paternity leave (2019). Retrieved from: http://www.index.go.kr/potal/stts/idxMain/selectPoSttsIdxSearch.do?idx_cd=1504&stts_cd=150401&freq=Y (website)
27. Jeong, S., McLean, G.N., Park, S.J.J.o.W.L.: Understanding informal learning in small-and medium-sized enterprises in South Korea (2018)
28. Herzberg, F., Bernard, M., Snyderman, B.B.: The Motivation to Work. Wiley, New York (1959)
29. Öncer, A.Z., Yıldız, M.L.J.P.-S., Sciences B: The impact of ethical climate on relationship between corporate reputation and organizational identification. Procedia. Soc. Behav. Sci. **58**, 714–723 (2012)
30. Wang, C., Xu, J., Zhang, T.C., Li, Q.M.J.J.o.H., Management T: Effects of professional identity on turnover intention in China's hotel employees: the mediating role of employee engagement and job satisfaction. J. Hosp. Tour. Manag. **45**, 10–22 (2020)
31. Mishra, S., Tripathy, H.K., Mallick, P.K., Bhoi, A.K., Barsocchi, P.: EAGA-MLP—an enhanced and adaptive hybrid classification model for diabetes diagnosis. Sensors. **20**(14), 4036 (2020)
32. Mishra, S., Mallick, P.K., Jena, L., Chae, G.S.: Optimization of skewed data using sampling-based preprocessing approach. Frontiers in Public Health. **8**, –274 (2020)
33. Mishra, S., Tripathy, H.K., Mishra, B.K.: Implementation of biologically motivated optimisation approach for tumour categorisation. Int. J. Comput. Aided Eng. Technol. **10**(3), 244–256 (2018)
34. Mishra, S., Mallick, P.K., Tripathy, H.K., Bhoi, A.K., González-Briones, A.: Performance evaluation of a proposed machine learning model for chronic disease datasets using an integrated attribute evaluator and an improved decision tree classifier. Appl. Sci. **10**(22), 8137 (2020)
35. Mallick, P.K., Mishra, S., Chae, G.S.: Digital media news categorization using Bernoulli document model for web content convergence. Pers. Ubiquit. Comput. (2020). https://doi.org/10.1007/s00779-020-01461-9

The Influence of Consultant Competency and Consulting Service Quality on Small-Medium Enterprise's Management Performance

Tae-Young Kim and Yen-Yoo You

1 Introduction

As modern society develops and changes at an incredible speed, unlike the past, the business environment of the company is also rapidly changing from the past. From the company's perspective, the role of management consulting is critical in predicting uncertain management conditions in advance and objectively diagnosing the current situation. Besides, this importance is not only for large companies but also for small- and medium-sized enterprises, and quality management consulting is essential. Therefore, this study aims to examine the relationship between the competency of consultants, the key factors in management consulting, the quality of SME management consulting services, and, finally, the management performance of SME. This study verified the critical qualities of consultants and quality factors of consulting services that SME perceived as necessary, and also revealed the relationship with management performance. In the future, it has practical implications for the management of small business consulting for SME.

2 Theoretical Considerations

2.1 SME Consulting Overview

It defined management consulting as "a service where independent qualified experts investigate problems related to corporate policies, organizations, procedures, methods, etc., recommend appropriate solutions, and at the same time assist in

T.-Y. Kim (✉) · Y.-Y. You
Department of Knowledge service & Consulting, Hansung University, Seoul, South Korea
e-mail: threey0818@hansung.ac.kr

S. R. Samanta et al. (eds.), *Cognitive Computing for Risk Management*,
EAI/Springer Innovations in Communication and Computing,
https://doi.org/10.1007/978-3-030-74517-2_10

improvement" [1]. Management consulting is "Managing organizational change to ask valuable questions to companies commissioned by consultants with knowledge and experience in management and to propose, advise and solve solutions" [2]. By doing this, it is an activity to build a network and receive money in return, and an activity that has a social responsibility of creating knowledge and sharing knowledge. The International Council of Management Consulting Institutes (ICMCI) defines management consulting as the act of providing independent advice and support for the management process to customers with management responsibilities and the subject of consultant independence and consulting. The concept is clearly defined. It goes through a particular training course to assist qualified people in identifying and analyzing business problems from an independent and objective standpoint under a contract with the customer and propose solutions to the customer, and consulting is defined as an advice service that provides help when customers asked for help with implementing solutions [3].

2.2 Consultation on the Capabilities of Consultants

Consultants are professionals who receive compensation based on their professional knowledge or experience and assist in accomplishing and guiding goals by analyzing business management and business problems or tasks and presenting advice and solutions. Due to the nature of these consultants, it verified that many previous studies had a significant effect on consultant competency, consulting performance, and customer satisfaction. There are various prior studies on the competency of consultants, but as a representative study, that affects essential parts of work performed by individuals, has high relevance to work performance, and is widely accepted by organizations. It defined as "a collection of knowledge, skills, and attitudes that can be measured against and improved through training and development" [4]. While presenting the factors necessary for successful consulting, it can analyze and solve problems with the competency required for consultants, "the ability to plan and operate schedules with limited resources such as consulting projectability, cost, and time Etc. He also emphasized that consultants should have new knowledge and original ideas" [5]. It cited pure consultant competency, expertise, and innate integrity as characteristics of a consultant. Here, "the intrinsic integrity was composed of the complicated concepts of motivation, ethics, objectivity, honesty, loyalty, and confidence" [6]. Similarly, in this study, "SERVQUAL composed of 5 dimensions of service quality: Reliability, Assurance, Tangibles, Empathy, and Responsiveness. The model defined as a management consulting service quality" [7].

2.3 *Management Consulting Service Quality Consideration*

Before looking at management consulting service quality, it is necessary to first look at service quality. A study defined "service quality as the difference between the customer's expectations of the service provided by the service company and the customer's perception of the service provided." [8]. It defined "service quality as the customer's overall attitude or judgment about the superiority or excellence of a particular service" [9]. It can measure the quality of service as the difference between the customer's expectation and perception and presented the SERVQUAL model as a multi-item measure to measure this. This measurement method measures the expected service level and the service recognition level at the same time. According to this model, service quality is measured by comparing the difference between the expected and perceived levels of customers for a given service [10]. Looking at the recent research related to consulting service quality, Rust and Zahorik proved that his excellent service quality has the effect of preventing the departure of existing customers and attracting new customers [11]. Parasuraman et al. showed in his research that "there is a positive correlation between the quality of service factor and the two items of the customer's intention to repurchase and the recommendation doctor" [7]. It was empirically analyzed that "all 12 service quality components have a positive effect on customer loyalty in a study on the effect of service quality on customer loyalty in E-banking" [12].

2.4 *Examining the Relationship Between Consultant Competency and Management Consulting Service Quality and Management Performance*

The professionalism and proficiency of the consultant, the customer's confidence, the customer's information needs, and information provision responsibilities, and other consultants' plans before the consultant and the client conduct consulting meets the needs of the consulting client. It said that the consulting result was affected by the confirmation process of identifying the type [13]. Consulting with the customer in the communication between the consultant and the customer, such as reporting "the progress of the consulting to the consulting performance factors, and providing the consultant's information, and promoting the project of the consultant, affects the consulting performance" [14].

3 Research Method

3.1 Research Problem

In this study, we tried to verify the specific factors of consultant competency and consulting service quality affecting the management performance of SME and to understand the relative influence of each factor. Accordingly, the following research problems are drawn up in this study. Research question 1 aimed to verify the influence of the three factors, such as competency, knowledge, and attitude, as a consultant competency, on the quality of consulting services for SME, and individually to examine the effect on reliability, confidence, and empathy. Research question 2 attempted to find out the effect of these three factors of consulting service quality on the management performance of SME.

Research Problem 1. How does consultant competency (competence, knowledge, attitude) affect the quality of SME consulting services?

1-1. How does consultant competency (competence, knowledge, attitude) affect the reliability of SME consulting services?
1-2. How does consultant competency (competence, knowledge, attitude) affect the confidence of SME consulting services?
1-3. How does the consultant's competency (competence, knowledge, attitude) affect SME consulting service empathy?

3.2 Research Method

This study conducted an online survey for 2 weeks from January 3 to January 17, 2020, for 230 domestic SME employees ($N = 230$) to verify the research problem. Among them, the final 223 copies are used for statistical analysis, excluding the outliers, unfaithful responses, and 7 nonresponses. For statistical verification, structural equation modeling is performed, and frequency analysis, Cronbach alpha reliability analysis, descriptive statistics analysis, and correlation analysis utilized SPSS (statistics package for social science) 21 statistical package and structured to verify the final research problem. Equation modeling verification was analyzed through AMOS (analysis of moment structures) 17 statistical package.

4 Research Results

4.1 Demographic Characteristics

The demographic characteristics of the respondents are as follows. Gender was relatively higher in male respondents than in female respondents, and the age range of respondents was in the order of 30s > 40s > 20s > 50s. SMEs currently employed have the highest number of working years 3–5 years, followed by 5–10 years, showing 3–10 years. The number of consulting experiences of SME currently employed was the highest in 3 times in the last 3 years, followed by 4 times >2 times >1 time > 5 times or more. The total number of employees was 223, and all respondents were currently employed by SME nationwide as shown in Table 1.

4.2 Validation and Reliability of Measurement Tools

In order to verify the validity and reliability of the measurement tools in this study, confirmatory factor analysis was conducted, and the validity criterion was factor loading, and AVE (average variance extracted) and construct reliability were examined [15–19]. As a result, the factor placement was more than 0.500 in all questions (0.456–0.906), and the AVE met the criteria of 0.400–0.500 (0.431–0.743). The construct reliability also met the criteria of 0.700 or higher (0.694–0.896) to ensure validity. Next, as a result of reliability verification, it was found that the Cronbach

Table 1 Demographic characteristics of participants

		Frequency (n)	Percentage (%)
Gender	Male	139	62.3
	Female	84	37.7
Age group	20s	40	17.9
	30s	105	47.1
	40s	64	28.7
	50s	13	5.8
Working years	Less than 3 years	21	9.4
	3–5 years	98	43.9
	5–10 years	84	37.7
	10–20 years	14	6.3
	More than 20 years	6	2.7
Consulting experience	1 time/3 years	34	15.2
	2 times/3 years	38	17.0
	3 times/3 years	68	30.5
	4 times/3 year	51	22.9
	More than 5 times/3 years	32	14.3
Total	223	100.0	

alpha value showed that all variables were above the reliability criterion of 0.600 (0.693–0.894), which ensured the validity and reliability of the measurement tool (Table 2).

The fit factor of the confirmatory factor analysis model is CMIN/DF = 1.738 (less than 3), IFI = 0.932, TLI = 0.919, CFI = 0.931 (excellent when over 0.90), RMSEA = 0.058 (it was found that the model suitability was less than 0.08).

4.3 Descriptive Statistics and Correlation Matrix

The results of the descriptive statistics of the measured variables are as follows. The average of the consultants' competencies was in the order of consultant ability > consultant knowledge > consultation attitude, and the quality of consulting services for SME was found in the order of empathy > reliability > reliability. As a result of the correlation between the measured variables, the reliability, certainty, and empathy of SME consulting service quality were found to have the highest correlation with the consultant attitude, and SME management performance was found to have the highest relationship with confidence in the consulting service quality factor (Table 3).

4.4 Results

The structural model results of structural equation modeling are shown in Table 4.

Research question 1 examined the effect of the competency, knowledge, and attitude of the consultant's competence on the reliability, conviction, and empathy of the SME consulting service quality. First, knowledge ($\beta = 0.277$, $p = 0.007$) and attitude ($\beta = 0.357$, $p = 0.003$) among the consultants' competency were influencing the reliability of SME consulting service quality ([RQ1-1]). Second, knowledge ($\beta = 0.235$, $p = 0.016$) and attitude ($\beta = 0.522$, $p = 0.000$) among consultants' competency were influencing the quality assurance of SME consulting services ([RQ1-2]). Lastly, three factors of competency ($\beta = 0.185$, $p = 0.053$), knowledge($\beta = 0.248$, $p = 0.001$), and attitude ($\beta = 0.450$, $p = 0.000$) among consultant competency factors are all SME. It was found that it had an impact on the quality sympathy of consulting services ([RQ1-3]).

Research question 2 examined the effect of reliability, certainty, and empathy, the quality factors of SME consulting service quality, on SME management performance. As a result of the study, it was found that two factors affect the business performance of SME: certainty ($\beta = 0.684$, $p = 0.000$) and empathy ($\beta = 0.208$, $p = 0.029$).

The structural model has good fits such as CMIN/DF = 1.907, IFI = 0.915, TLI = 0.901, CFI = 0.913, and RMSEA = 0.064.

Table 2 The results of validity and reliability

Variables	Items	β	B	S.E.	t	AVE	C.R.	Cronbach alpha
Consultant skills	Consultants have the ability to diagnose corporate problems	0.681	1			0.493	0.871	0.872
	The consultant has the ability to analyze the problems found in the company	0.748	1.1	0.091	12.086***			
	The consultant has the ability to analyze problems and present alternatives to the results of diagnosis	0.711	1.052	0.112	9.421***			
	Consultants have the ability to manage the entire consulting process	0.804	1.135	0.109	1.458***			
	Consultants have the ability to collect information related to consulting	0.633	0.975	0.115	8.5***			
	The consultant has the communication skills necessary for smooth consulting	0.62	0.922	0.111	8.333***			
	Consultants have the ability to predict the future to adapt to change	0.701	1.163	0.125	9.331***			
Consultant knowledge	The consultant has knowledge of the general management necessary to conduct consulting	0.784	1			0.539	0.817	0.789
	The consultant has the expertise necessary for consulting	0.848	1.025	0.08	12.847***			
	The consultant has knowledge of organizational theory and organizational management	0.784	0.876	0.074	11.896***			
	There is empirical knowledge (know-how) that the consultant has gained from performing various consulting	0.456	0.651	0.099	6.552***			

(continued)

Table 2 (continued)

Variables	Items	β	B	S.E.	t	AVE	C.R.	Cronbach alpha
Consultant attitude	No false results were presented or exaggerated when consulting was conducted	0.865	1			0.634	0.873	0.891
	Consulting complied with the scope and scope of work	0.843	1.025	0.067	15.259***			
	Responsible for the things that happen in the consulting process	0.678	0.763	0.069	11.109***			
	No additional charges were required other than those specified in the contract	0.786	0.843	0.061	13.789***			
Consulting reliability	The schedule for consulting was followed as planned	0.731	1			0.455	0.709	0.695
	There was consistency and accuracy in the information or knowledge provided by the consulting	0.752	0.909	0.101	8.993***			
	Consulting kept and kept company information and company secrets	0.514	0.697	0.105	6.638***			
Consulting confidence	I was able to entrust consulting tasks with confidence in consulting requests	0.606	1			0.446	0.706	0.704
	Consulting has come to believe in the skills of an expert with knowledge in a specialized field	0.655	1.118	0.154	7.281***			
	I was convinced of the stable operation and performance of the consulting project	0.737	1.285	0.165	7.787***			

(continued)

Table 2 (continued)

Variables	Items	β	B	S.E.	t	AVE	C.R.	Cronbach alpha
Consulting empathy	Consulting has sincerely considered the company's interests	0.906	1			0.743	0.896	0.894
	Consulting gave the company individual attention and human consideration	0.876	1.032	0.057	18.064***			
	Consulting actively identified and recognized the company's needs	0.8	0.891	0.058	15.434***			
Management performance	Through the consulting, customer satisfaction was improved	0.645	1			0.431	0.694	0.693
	Through consulting, the company's business process has been improved	0.637	0.881	0.111	7.947***			
	Through consulting, the problem solving and work performance of company members improved	0.687	0.901	0.107	8.435***			

Note. ***$p < 0.001$

5 Conclusion

Today, management consulting is becoming more critical for small- and medium-sized enterprises (SMEs), rather than just consulting with large companies according to various market environments and rapidly changing business environments. Therefore, this study examined the relationship between the competency of consultants, the key factors in management consulting, the quality of SME management consulting services, and, finally, the performance of SME management by conducting a structural equation. As a result of the study, knowledge and attitude among the competency of consultants showed that the reliability and confidence of SME consulting service quality increased, and in the case of empathy, all three factors are influenced by the attitude, ability, and knowledge of consultant competency. On the other hand, two factors of confidence and empathy among SME consulting service quality were found to improve SME management performance.

Based on this research results, the following conclusions can be drawn. First, based on the research results that knowledge and attitude two factors among

Table 3 Correlation analysis and descriptive analysis

	Consultant skills	Consultant knowledge	Consultant attitude	Consulting reliability	Consulting confidence	Consulting empathy	Management performance	*M*	*SD*
Consultant skills	1	0.546**	0.605**	0.430**	0.445**	0.585**	0.580**	3.920	0.768
Consultant knowledge		1	0.512**	0.428**	0.447**	0.535**	0.627**	3.624	0.783
Consultant attitude			1	0.447**	0.528**	0.640**	0.526**	3.590	0.904
Consulting reliability				1	0.476**	0.531**	0.418**	3.248	1.005
Consulting confidence					1	0.452**	0.552**	2.839	0.923
Consulting empathy						1	0.515**	3.848	0.996
Management performance							1	3.788	0.777

Note: $^{**}p < 0.01$

Table 4 The main results of research questions

			β	B	$S.E.$	t	p
Consultant skills	→	Consulting reliability	0.138	0.192	0.174	1.105	0.269
Consultant knowledge	→	Consulting reliability	0.277	0.331	0.123	2.691	0.007
Consultant attitude	→	Consulting reliability	0.357	0.372	0.126	2.958	0.003
Consultant skills	→	Consulting confidence	0.099	0.094	0.111	0.852	0.394
Consultant knowledge	→	Consulting confidence	0.235	0.192	0.079	2.419	0.016
Consultant attitude	→	Consulting confidence	0.522	0.371	0.088	4.211	***
Consultant skills	→	Consulting empathy	0.185	0.263	0.136	1.933	0.053
Consultant knowledge	→	Consulting empathy	0.248	0.303	0.095	3.186	0.001
Consultant attitude	→	Consulting empathy	0.45	0.48	0.098	4.873	***
Consulting reliability	→	Management performance	0.105	0.079	0.072	1.108	0.268
Consulting confidence	→	Management performance	0.684	0.757	0.156	4.845	***
Consulting empathy	→	Management performance	0.208	0.154	0.07	2.185	0.029

*$p<0.05$, **$p<0.01$, ***$p<0.001$

consultant competency have a positive effect on SME consulting service quality, when conducting management consulting for small- and medium-sized enterprises, specialized knowledge and client among consultants' qualities. It is thought that the quality and quality of the consultant's attitude and attitude should be managed and promoted. Second, among the quality of consulting services for SME, it was found that conviction and empathy ultimately improve SME management performance. Notably, the influence of confidence is considerable. It can be inferred that confidence has a high impact on the management performance of SME. In addition, since confidence is a factor that is formed based on trust and trust between a consultant and a client SME, it is necessary to focus on consulting preparation and operation management to build confidence based on trust and trust in consulting.

This study verified the critical qualities of consultants and quality factors of consulting services that SEMs perceived as necessary, and also revealed the relationship with management performance. In the future, it has practical implications for the management of small business consulting for SME. I think it can be presented. Besides, in the future, considering the characteristics of various small- and medium-sized enterprises, if a related study is conducted for a large number of people, it is expected that more logical and reasonable results can be presented.

Acknowledgments This study was financially supported by Hansung University.

References

1. Kubr, M.: How to select and use consultants, 2nd. ILO, Geneva (1997)
2. Iansiti, M., Levien, R.: The Keystone Advantage: What the New Dynamics of Business Ecosystems Mean for Strategy, Innovation, and Sustainability. Harvard Business Press, Boston (2004)

3. Greiner, L.E., Metzger, R.O.: Consulting to Management. Prentice-Hall, Englewood Cliffs (1983)
4. Parry, S.B.: Just what is a competency? (and why should you care?). Training. **35**(6), 58 (1996)
5. Rynning, M.: Successful consulting with small and medium-sized vs. large clients: meeting the needs of the client? Int. Small Bus. J. **11**(1), 47–60 (1992)
6. McLachlin, R.D.: Factors for consulting engagement success. Manag. Decis. **37**(5), 394–404 (1999)
7. Parasuraman, A., Zeithaml, V.A., Berry, L.L.: The behavioral consequences of service quality. J. Mark. **60**(2), 31–46 (1996)
8. Parasuraman, A., Zeithaml, V.A., Berry, L.L.: SERVQUAL: a multiple-item scale for managing consumer perceptions of service quality. J. Mark. **18**(45), 437–449 (1998)
9. Parasuraman, A.: Customer-oriented corporate cultures are crucial to services marketing success. J. Serv. Mark. **1**(1), 39–46 (1987)
10. Parasuraman, A., Zeithaml, V.A., Berry, L.L.: Servqual: a multiple-item scale for measuring consumer perc. J. Retail. **64**(1), 12 (1988)
11. Rust, R.T., Zahorik, A.J.: Customer satisfaction, customer retention, and market share. J. Retail. **69**(2), 193–215 (1993)
12. Kaur, H., Allan, E.L., Mamadu, I., Hall, Z., Ibe, O., El Sherbiny, M., et al.: Quality of artemisinin-based combination formulations for malaria treatment: prevalence and risk factors for poor quality medicines in public facilities and private sector drug outlets in Enugu, Nigeria. PLoS One. **10**(5), e0125577 (2015)
13. Adamson, I.: Management consultant meets a potential client for the first time: the pre-entry phase of consultancy in SMEs and the issues of qualitative research methodology. Qual. Mark. Res. **3**(1), 17–26 (2000)
14. Boswell, P., Walker, L.: Procurement and Process Design. FIDIC and Lorna Walker Consulting Ltd, Geneva, London (2004)
15. Mishra, S., Tripathy, H.K., Mallick, P.K., Bhoi, A.K., Barsocchi, P.: EAGA-MLP—an enhanced and adaptive hybrid classification model for diabetes diagnosis. Sensors. **20**(14), 4036 (2020)
16. Mishra, S., Mallick, P.K., Jena, L., Chae, G.S.: Optimization of skewed data using sampling-based preprocessing approach. Front. Public Health. **8**, 274 (2020)
17. Mishra, S., Tripathy, H.K., Mishra, B.K.: Implementation of biologically motivated optimisation approach for tumour categorisation. Int. J. Comput. Aided Eng. Technol. **10**(3), 244–256 (2018)
18. Mishra, S., Mallick, P.K., Tripathy, H.K., Bhoi, A.K., González-Briones, A.: Performance evaluation of a proposed machine learning model for chronic disease datasets using an integrated attribute evaluator and an improved decision tree classifier. Appl. Sci. **10**(22), 8137 (2020)
19. Mallick, P.K., Mishra, S., Chae, G.S.: Digital media news categorization using Bernoulli document model for web content convergence. Pers. Ubiquit. Comput. (2020). https://doi.org/10.1007/s00779-020-01461-9

Index

S. R. Samanta et al. (eds.), *Cognitive Computing for Risk Management*,
EAI/Springer Innovations in Communication and Computing,
https://doi.org/10.1007/978-3-030-74517-2

The manufacturer's authorised representative in the EU is Springer Nature Customer Service Centre GmbH, Europaplatz 3, 69115 Heidelberg, Germany. If you have any concerns regarding our products, please contact ProductSafety@springernature.com

Printed and bound by CPI Group (UK) Ltd, Croydon, CR0 4YY

30/09/2026

02225306-0012